Teubner Studienbücher

Informatik

Ehrig u. a.: **Universal Theory of Automata**
240 Seiten. DM 22,80

Hotz: **Informatik: Rechenanlagen**
Struktur und Entwurf, 136 Seiten. DM 14,80 (LAMM)

Kandzia/Langmaack: **Informatik: Programmierung**
234 Seiten. DM 18,80 (LAMM)

Maurer: **Datenstrukturen und Programmierverfahren**
222 Seiten. DM 25,80 (LAMM)

Schnorr: **Rekursive Funktionen und ihre Komplexität**
191 Seiten. DM 24,80 (LAMM)

Wirth: **Systematisches Programmieren**
Eine Einführung. 160 Seiten. DM 14,80 (LAMM)

Mathematik

Böhmer: **Spline-Funktionen**
Theorie und Anwendungen. 340 Seiten. DM 24,80

Clegg: **Variationsrechnung**
138 Seiten. DM 14,80

Collatz: **Differentialgleichungen**
Eine Einführung unter besonderer Berücksichtigung der Anwendungen.
5. Aufl. 226 Seiten. DM 18,80 (LAMM)

Collatz/Krabs: **Approximationstheorie**
Tschebyscheffsche Approximation mit Anwendungen. 208 Seiten. DM 26,80

Constantinescu: **Distributionen und ihre Anwendung in der Physik**
144 Seiten. DM 16,80

Fischer/Sacher: **Einführung in die Algebra**
238 Seiten. DM 15,80

Grigorieff: **Numerik gewöhnlicher Differentialgleichungen**
Band 1: Einschrittverfahren. 202 Seiten. DM 13,80
Band 2: Mehrschrittverfahren

Hainzl: **Mathematik für Naturwissenschaftler**
311 Seiten. DM 29,– (LAMM)

Hilbert: **Grundlagen der Geometrie**
11. Aufl. VII, 271 Seiten, DM 18,80

Kochendörffer: **Determinanten und Matrizen**
IV, 148 Seiten. DM 14,80

Stiefel: **Einführung in die numerische Mathematik**
Eine Darstellung unter Betonung des algorithmischen Standpunktes
4. Aufl. 257 Seiten. DM 18,80 (LAMM)

Fortsetzung auf der 3. Umschlagseite

Universal Theory of Automata

A Categorical Approach

by Dr. rer. nat. H. Ehrig
cand. math. K.-D. Kiermeier
Dipl.-Math. H.-J. Kreowski
and Dipl.-Math. W. Kühnel

Technische Universität Berlin

1974. With numerous figures and examples

B. G. Teubner Stuttgart

Dr. rer. nat. Hartmut Ehrig

1944   born in Angermünde, Germany
1963-1969   study of mathematics, physics and
              theoretical informatics at the Tech-
              nische Universität Berlin (TUB)
1969   Dipl.-Math.
1971   Dr. rer. nat.
1970-1972   Wissenschaftlicher Assistent at the
              Fachbereich Mathematik of the TUB
since 1972   Assistenzprofessor at the Fachbereich
              Kybernetik of the TUB
1974   venia legendi

Klaus-Dieter Kiermeier

1949   born in Berlin, Germany
since 1969   study of mathematics, physics and
              theoretical informatics at the TUB

Dipl.-Math. Hans-Jörg Kreowski

1949   born in Berlin, Germany
1969-1974   study of mathematics, economics and
              theoretical informatics at the TUB
1974   Dipl.-Math.
since 1974   Wissenschaftlicher Assistent at the
              Fachbereich Kybernetik of the TUB

Dipl.-Math. Wolfgang Kühnel

1950   born in Berlin, Germany
1969-1974   study of mathematics, economics and
              theoretical informatics at the TUB
1974   Dipl.-Math.

ISBN 978-3-519-02054-7   ISBN 978-3-322-96644-5 (eBook)
DOI 10.1007/978-3-322-96644-5

Printer: J. Beltz, Hemsbach/Bergstr.

Cover design: W. Koch, Sindelfingen

Preface

Our purpose in writing this book is to present a universal
theory of automata which on one hand unifies the theories of
several well-known types of automata and on the other hand
allows interesting new applications and results. The framework for our development is category theory, especially
universal constructions in monoidal categories. But we will
carefully motivate and introduce all those (and only those)
notions and results of category theory which are needed in
our approach. The reader is only assumed to be familiar with
sets, deterministic functions, relations and the basic notions of structural mathematics. However, some knowledge of
discrete probability distributions, linear algebra and
general topology would be useful in understanding the corresponding applications and in having a better background for
the general theory. All our constructions and results are
motivated and interpreted carefully with respect to the
classical theory of deterministic, partial, linear, topological, nondeterministic, relational and stochastic automata.

The book is mainly devoted to students of theoretical computer science or mathematics and can be used as a textbook
in graduate courses or seminars. On the other hand it will
also be useful for many other people, who are concerned
with the interesting new research area of category theory
applied to computation and control.

This book is based on the report [28] of a research seminar
in 1972/73, which was also used in an advanced course of
automata theory in 1973, and on other research articles of
the authors [25-32,57]. Moreover we were influenced by the
following papers on categorical automata theory [4,5,6,17,33,
44,45,46,60,78] and by several other books and articles on
category or automata theory, above all [2,14,20,23,24,50,54,
56,59,64,71,75].

For several helpful discussions and suggestions we are
grateful to many colleagues and students, especially to
M. A. Arbib, W. Brauer, S. Eilenberg, J. A. Goguen, G. Hotz,
E. G. Manes and M. Pfender.
Special thanks are due to U. Brödner and W. Werner for ex-
cellent drawings of diagrams and figures, and H. Barnewitz,
G. Ehrig, B. Mahr and W. Merzenich for proof-reading and
useful comments concerning the manuscript.
Finally, we wish to thank the Teubner-Verlag for friendly
co-operation and quick publication.

Berlin, July 1974   H. Ehrig

                                           K.-D. Kiermeier

                                           H.-J. Kreowski

                                           W. Kühnel

Contents

Introduction

In the development of abstract automata theory numerous
structures have been studied using different terminology and
methods but arriving mostly at similar results. For deterministic, partial, linear and some types of topological automata for example, it is possible to construct equivalent minimal automata which are uniquely determined up to isomorphism.
On the other hand the different uncanonical constructions
for the minimization of nondeterministic, relational and
stochastic automata can be seen under a common point of view.

The main purpose of this book is to present a unified description and to develop, as far as possible, the main lines
of a common theory for all these different types of automata.
Furthermore we will introduce new techniques in our general
approach leading to canonical proofs and new results in several special cases. Finally we will show that our theory is
applicable to interesting new examples which have not been
studied in classical automata theory before.

Let us begin with the unified description. A deterministic
automaton A consists of sets I, O, S, called input, output and states respectively, and functions $d:S \times I \to S$ and
$l:S \times I \to O$ assigning to each state and input the next state
and the output respectively. Since functions in our sense
are always deterministic, next state and output are uniquely
determined by state and input. If we wish to have several
possibilities for the next state and output we have to replace d and l by nondeterministic functions from the cartesian product $S \times I$ to S and O respectively. This leads
to the notion of a nondeterministic automaton A . Moreover
if we have given probability distributions for the transitions to the next state and output we have stochastic chan-
nels $d:S \times I \to S$ and $l:S \times I \to O$ and thus a stochastic automaton. Finally we get the notions of linear and topological

automata if we replace the sets I, O, S by vector or topological spaces and d, l by linear or continuous functions
respectively. Thus we get the following table without the
bottom row for the moment:

I, O, S	d, l	automaton type
sets	deterministic functions	deterministic
sets	nondeterministic functions	nondeterministic
sets	stochastic channels	stochastic
vector spaces	linear functions	linear
topological spaces	continuous functions	topological
objects	morphisms	automaton in a category

Now it is natural to ask whether there are common general
notions for I, O, S and d, l which can be specialized
respectively to the above meanings. In fact I, O, S and
d, l can be viewed as "objects" and "morphisms" in a suitable category $\underline{K}$. These are exactly the constituent parts of
a category. Roughly spoken a category $\underline{K}$ is a collection of
objects A, B, ... and morphisms $f:A \to B$, ... together
with an associative composition rule assigning to each pair
of morphisms $f:A \to B$ and $g:B \to C$ a composite morphism
$g \circ f:A \to C$. Sets and deterministic functions for example
constitute the category $\underline{Set}$ of sets, and vector spaces
over a fixed field F together with F-linear functions define the category $\underline{V}$ of vector spaces. In a similar way the
other examples lead to the categories $\underline{ND}$, $\underline{Stoch}$ and $\underline{Top}$ of
nondeterministic functions, stochastic channels and topologi-
cal spaces respectively.

Thus our desired unified description is given by the notion
of an automaton in a category $\underline{K}$, or more precisely in a
monoidal category $(\underline{K}, \otimes)$. Such an automaton consists of

objects I, O, S and morphisms d, l in the category $\underline{K}$ of
the form

$$O \xleftarrow{\qquad l \qquad} S \otimes I \xrightarrow{\qquad d \qquad} S$$

where $\otimes$ is a suitable generalization of the cartesian product $\times$. A more detailed version of these ideas together
with exact definitions and further examples is given in chapter 1 .

The next step is to ask for the subjects in automata theory
which can be treated in our general framework. Up to now
categorical methods have been applied to the following
fields:

> 1. Decomposition and Synthesis
>
> 2. Reduction and Minimization
>
> 3. Behavior and Realization
>
> 4. Structure Theory.

A detailed categorical investigation of subject 1 is given
in [18] and thus we can afford to omit this field here.
Our main purpose is to study subjects 2 and 3 for automata in monoidal categories based on [17,25-32,44-46] and to
interpret the constructions and results in terms of classical automata theory. For other approaches in this area we
refer to [4,5,6,8,19,61]. Finally subject 4 - structure
theory of automata - has been studied in [33] for deterministic automata in great detail, and will be extended to
several other types in this book.

The problems of reduction, minimization and realization are
motivated in chapter 2 . The constructions and results in
the case of deterministic automata are given in such a way
that they can be generalized to automata in monoidal categories of deterministic type in chapters 4 and 5 and, with
some modifications, also to automata of nondeterministic
type. Basic notions of automata and system theory like
"behavior", "minimal", "reduced", "realizable" and "equiva-

lence" are formulated in great generality and compared with each other in chapter 3 . In particular we discuss the "Minimal Realization Principle" which was stated by J. A. G o g u e n in [44,45]. We show that several interesting corollaries can be deduced from that principle which can be applied in the following chapters. Moreover it is possible to classify the different constructions for reduction and minimization of deterministic and nondeterministic type automata respectively. Applications of this approach to several other examples in automata and system theory are given in [30].

In chapter 4 we start with the general theory for automata of deterministic type which can, for example, be applied to deterministic, partial, bilinear and topological automata. The essential point is that we only use universal properties of morphisms instead of a description by elements. This procedure is typical for categorical approaches. Let us explain by the following important example how to come from a description by elements to a universal one:

The output function $1:S \times I \to 0$ of a deterministic automaton with next state function $d:S \times I \to S$ can be extended to input strings $x_1 x_2 \ldots x_n$, which are elements of the free semigroup I^+ over I , in the following way: For each state s the last output $1^+(s, x_1 x_2 \ldots x_n)$, which is an element in 0 , is recursively defined by

$$1^+(s, x_1) = 1(s, x_1) \quad \text{and}$$
$$1^+(s, x_1 x_2 \ldots x_n) = 1^+(d(s, x_1), x_2 \ldots x_n) \ .$$

Thus we get the extended output function $1^+:S \times I^+ \to 0$ which leads to the machine function $M:S \to \langle I^+, 0 \rangle$, where $\langle I^+, 0 \rangle$ is the set of all functions from I^+ to 0 . M assigns to each state s the corresponding input-output behavior $M(s):I^+ \to 0$ defined by

$$M(s)(w) = 1^+(s, w) \quad \text{for all strings } w = x_1 \ldots x_n.$$

These constructions of the functions 1^+ and M are typical

for a description by elements. Now we want to consider a
universal construction of M which can also be generalized
to several other types of automata. Given the function
$1^+:S\times I^+ \to 0$ the universal construction of $M:S \to <I^+,0>$ is
based on the following lemma which is used as an axiom in
our theory.

<u>Lemma 1</u>: For each pair of sets T and 0 there is a set
$<T,0>$ and a function $ev:<T,0>\times T \to 0$, called evaluation,
having the following universal property: For each function
$f:S\times T \to 0$ there is a unique function $\bar{f}:S \to <T,0>$ such
that the following diagram is commutative, which means
$ev \circ (\bar{f}\times id_T) = f$.

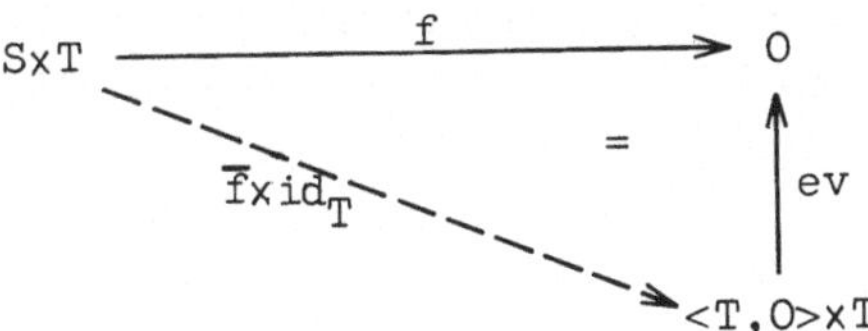

<u>Proof</u>: Let $<T,0>$ be the set of all functions $g:T \to 0$ and
$ev:<T,0>\times T \to 0$ is defined by $ev(g,t) = g(t)$ for all g in
$<T,0>$ and t in T . Now given $\bar{f}:S \to <T,0>$ with
$ev \circ (\bar{f}\times id_T) = f$ we have for all s in S and t in T :

$$f(s,t) = ev \circ (\bar{f}\times id_T)(s,t) = ev(\bar{f}(s),t) = \bar{f}(s)(t)$$

and thus $\bar{f}$ is uniquely determined by f . Conversely, defining $\bar{f}$ by the above equation, the diagram commutes. ∎

Thus given $1^+:S\times I^+ \to 0$ we get a unique function M satisfying $ev \circ (M\times id_{I^+}) = 1^+$ which coincides with our machine
function defined above elementwise. In fact, lemma 1, which
is formulated for the category of sets, is also true, in a
corresponding version, for the categories of vector spaces
and topological spaces. In our general approach, lemma 1
will be an axiom for our monoidal category $(\underline{K},\otimes)$ if we re-

place sets, functions and × by $\underline{K}$-objects, $\underline{K}$-morphisms and
⊗ respectively. $(\underline{K},⊗)$ is called closed category in this
case. It will be shown in chapters 4 and 5 that most of the
constructions and results concerning behavior characterization, reduction and minimization of deterministic automata
can be generalized to automata in closed categories. These
results can be applied to automata of deterministic type
like deterministic, partial and several sorts of linear and
topological automata.

On the other hand lemma 1 is not true for nondeterministic
functions for example, such that a new concept is necessary
for automata of nondeterministic type like nondeterministic
and stochastic automata. The basic idea in treating this case
is to consider each nondeterministic function $f:A → B$ as a
deterministic one from A to PB where PB is the powerset
of B without the empty subset. Let $v:PB → B$ be the nondeterministic function assigning to each non-empty subset
B' of B all elements of B' then we have:

<u>Lemma 2</u>: $v:PB → B$ has the following universal property:
For each nondeterministic function $f:A → B$ there is a
unique deterministic function $f':A → PB$ such that the following diagram is commutative, i.e. $v∘f' = f$.

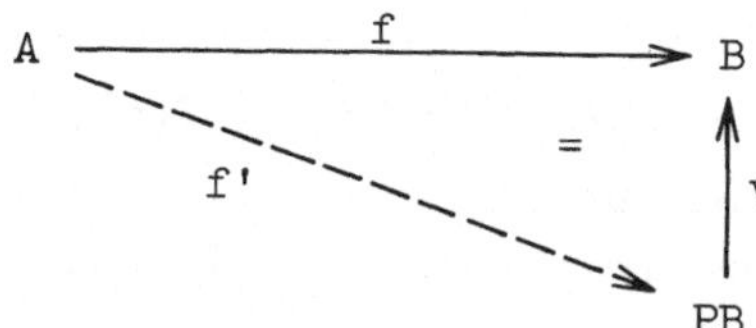

<u>Proof</u>: Given $f':A → PB$ with $v∘f' = f$ we have for each a∈A
$$f(a) = v∘f'(a) = v(f'(a)) = f'(a) .$$

Thus f' is uniquely determined by f . Vice versa the above
diagram commutes for $f':A → PB$ defined by this equation. ∎

In a similar way each stochastic channel f:A → B can be regarded as a deterministic function f':A → PB where PB is the set of all probability distributions on B . In our general approach of automata in a monoidal category (K,⊗) we assume that lemma 2 is true if we replace nondeterministic functions by morphisms in K and deterministic ones by morphisms in a suitable closed subcategory K' of K . Roughly spoken such a category (K,⊗) will be called pseudoclosed relative (K',⊗) and there are several other interesting examples for this situation, which, in addition, allow the treatment of relational and several types of relational topological automata. In chapters 6 and 7 it will be shown that the classical constructions for reduction and minimization of nondeterministic and stochastic automata can be generalized to automata in pseudoclosed categories. For the construction of reduced automata we can use a "cointersection", which is a well-known categorical concept, and it is interesting that proofs become much simpler in this way than they were in the classical theories (cf. [20]). Moreover we have some new results, concerning the behavior characterization and finite realizations, and several applications of the general theory in chapter 3 which can be used to classify the different types of minimal and reduced automata in the nondeterministic case.

In chapter 8 we construct for each automaton of nondeterministic type a corresponding one of deterministic type and discuss their relationships. This is a generalization of the construction of power automata which is useful in reducing problems for nondeterministic to those for deterministic type automata which are, of course, much better known.

All the constructions given in chapters 4 to 8 are extended in chapter 9 to automata with fixed initial state. Moreover we give constructions for reachable subautomata, i.e. each state can be reached from the initial state by an input string, and discuss the compatibility with reduction and minimization. An important concept for the study of initial automata are free and minimal realizations.

An initial automaton in a closed category turns out to be minimal if it is reachable and observable, i.e. different states have different input-output behavior. Unfortunately this is no longer true for nondeterministic and stochastic automata for example. Moreover there is no general construction to get an equivalent initial automaton with minimal number of states in these examples. But we are able to present a new concept, called scoop minimization, in chapter 10 which allows the replacement of states by equivalent subsets of the remaining states. This concept seems to be a rather good general approximation for the minimization problem. The scoop construction is formulated for automata in pseudo-closed categories and can be applied to nondeterministic, relational and stochastic automata.

As mentioned before, a general structure theory of automata in monoidal categories is developed in chapter 11. We discuss especially existence and construction of products, coproducts, equalizers, coequalizers and free constructions and we give interpretations for our examples.

Because we do not want to start with all the basic categorical notions which are needed for the theory, these concepts are introduced in the text only informally or in special cases. But we give references to the appendix in chapter 12 where we have summarized the corresponding exact categorical definitions and results. Moreover we have a list of special symbols, a subject index and a list of references at the end of the book.

Finally let us remark that this book is only an introduction to a universal theory of automata in categorical terminology and that there are several other results which we have not mentioned and a lot of more interesting subjects which should be studied in that framework, but which have not been worked out in detail up to now.

1. Unified Representation of Automata

Starting with the examples of deterministic, nondeterministic, stochastic, linear, bilinear and topological automata
it will be shown that the notion of automata in monoidal categories provides an appropriate general setting to get a
unified representation of all these different kinds of automata. Moreover partial, relational and relational topological
automata are given as examples and several other types will
be mentioned in the following chapters. For each special
type we give references to the literature. The idea of stud-
ying automata in monoidal categories is due to L. B u d a c h
and H.-J. H o e h n k e in [17] and J. A. G o g u e n in
[44,45]. We start with the definition and automata theoretic
interpretation of automata in the sense of G. H. M e a l y.
The notions of Moore- and Medvedev-automata will be given at
the end of the chapter.

<u>1.1 Definition</u> (D e t e r m i n i s t i c   A u t o -
m a t a) : A <u>deterministic automaton</u> is a 5-tuple
$A = (I,O,S,d,l)$ with sets
   I ("input symbols" or "input")
   O ("output symbols" or "output")
   S ("states")   and functions
   $d: SxI \to S$ ("state transition function") and
   $l: SxI \to O$ ("output function")
where SxI is the cartesian product of S and I. A function in
our sense is always deterministic and totally defined unless
otherwise specified.

The theory of deterministic automata can be found in most of
the books on automata theory, e.g. [2,33,37,38,39,49,54,62,
75].

<u>1.2 Processing of Input Strings</u>: In order to get the processing of input strings instead of input symbols we have to extend $d:S\times I \to S$ and $l:S\times I \to O$ to the free monoid $I*$ and the free semigroup I^+ on I respectively. Thus we get functions $d*:S\times I* \to S$ and $l^+:S\times I^+ \to O$ defined by recursion for all $s\in S$, $x\in I$, $w\in I*$:

$$d*(s,\square) = s \qquad\qquad (\square \in I* \text{ empty string})$$
$$d*(s,xw) = d*(d(s,x),w)$$
$$l^+(s,x) = l(s,x)$$
$$l^+(s,xw) = l^+(d(s,x),w)$$

It is easy to show that l^+ has in addition the representation $l^+(s,wx) = l(d*(s,w),x)$.

Starting with a state $s\in S$ and a symbol $x\in I$ the next state of the automaton is defined to be $d(s,x)\in S$ whereas the state transition corresponding to a sequence $x_1,\ldots,x_n$ of input symbols, i.e. a string $w = x_1\ldots x_n \in I*$, is given by $d*(s,w)\in S$. Similarly $l(s,x)\in O$ is the output of the automaton caused by $x\in I$ in the state $s\in S$ and $l^+(s,w)\in O$ is the last output symbol corresponding to the input string $w\in I^+$. Of course l also can be extended to $l*:S\times I* \to O*$ yielding the whole output string $y_1\ldots y_n \in O*$. The following picture (1.2.1) is a black box model of an automaton A when the substring $x_1\ldots x_k$ has been processed with $y_n = l^+(s,x_1\ldots x_n)$ $(n = 1,\ldots,k)$.

$$(1.2.1) \quad \xleftarrow{\quad y_1 \;\cdots\; y_k \quad} \boxed{\; d*(s,x_1 \;\cdots\; x_k) \;} \xleftarrow{\quad x_{k+1} \;\cdots\; x_n \quad}$$

In view of technical interpretations it would be better to regard an automaton of the form $A* = (I*,O*,S,d*,l*)$ but for our theory it suffices to take $A = (I,O,S,d,l)$ which is in fact a generating system for $A*$.

<u>1.3 Examples</u>: We want to give two simple examples:
1. The deterministic automaton $A_1 = (I,O,S_1,d_1,l_1)$ given by $I = \{x\}$, $O = \{y_1,y_2\}$, $S_1 = \{1,2,3\}$,
$d_1(i,x) = i+1$ for $i = 1,2$ and $d_1(3,x) = 2$,

$l_1(i,x) = y_1$ for $i = 1,3$ and $l_1(2,x) = y_2$
can be represented as a (node-)labeled graph in the following way:

(1.3.1)

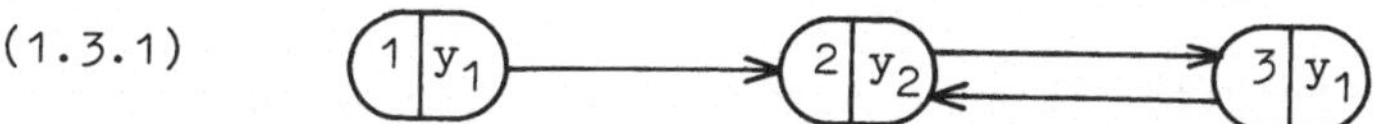

In our graph representations of automata with a single input
symbol $x \in I$ nodes and edges correspond to states and state
transitions respectively. The labels in the nodes indicate
the numbers of the states $i \in S$ and the corresponding output
$l(s,x) \in 0$.
Calculation of l^+ yields: $l_1^+(1,x^n) = l_1^+(2,x^{n+1}) = l_1^+(3,x^n) = y_1$
if n is odd and we get y_2 if n is even. Regarding the chains
with source i for $i = 1,2,3$ and length $n-1$ in the representing graph $l_1^+(i,x^n)$ is given by the label of the target.

2. Now we consider the graph

(1.3.2)

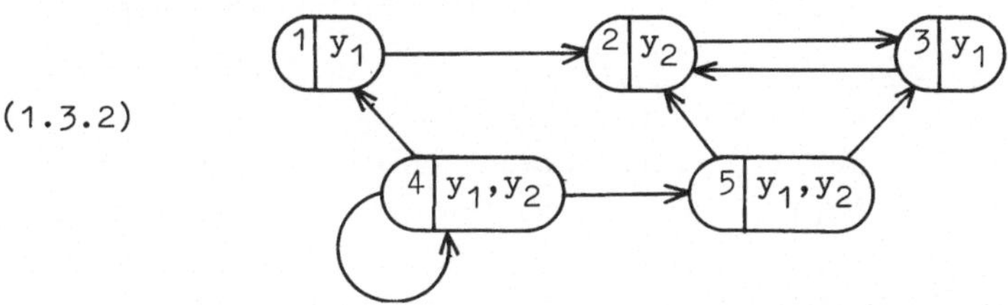

corresponding to an automaton $A_2 = (I,0,S_2,d_2,l_2)$ defined by
$S_2 = \{1,2,3,4,5\}$, $d_2(i,x) = d_1(i,x)$ and $l_2(i,x) = l_1(i,x)$ for
$i = 1,2,3$, $d_2(4,x) = \{1,4,5\}$, $d_2(5,x) = \{2,3\}$ and $l_2(4,x) =$
$= l_2(5,x) = \{y_1,y_2\}$. Since $d_2:S_2 \times I \to S_2$ and $l_2:S_2 \times I \to 0$ are
nondeterministic functions, A_2 is an example of a nondeter-
ministic automaton within the terms of the following definition 1.4:

1.4 Definition (N o n d e t e r m i n i s t i c   A u t o -
m a t a) : A nondeterministic automaton $A = (I,0,S,d,l)$
consists of sets I, 0, S and nondeterministic functions
$d:S \times I \to S$ and $l:S \times I \to 0$ with meaning similar to that
given in 1.1.

A nondeterministic function f:X → Y assigns to each x∈X
a non-empty subset f(x) of Y . Hence f can be regarded
as a function from X to ℘'O , i.e. the power set ℘O of
O without the empty subset.

Remark: In the deterministic case d and l together can
be regarded as one function t:SxI → SxO defined by t(s,x) =
= (d(s,x),l(s,x)) and vice versa d and l are uniquely
defined by such a t . This correspondence does not remain
true for nondeterministic functions d and l . In fact
both notions of nondeterministic automata are studied in the
literature, e.g. [12,38,49,75]. The same situation is given
in the case of partial, relational and stochastic automata.

1.5 Definition: (S t o c h a s t i c   A u t o m a t a) :
A stochastic automaton A = (I,O,S,d,l) again consists of
sets I, O, S but d:SxI → S and l:SxI → O are now
"(discrete) stochastic channels" in the following sense:

A (discrete) stochastic channel f:X → Y is defined as a
function f:X → <Y,[0,1]> assigning to each x∈X a
"(discrete) probability distribution" p = f(x) , which is a
function p:Y → [0,1] from Y to the unit interval satisfying
   (i)   p(y) ≠ 0 for at most a countable number of y∈Y
   (ii) $\sum_{y \in Y} p(y) = 1$.

Remark: In general <A,B> will denote the set of all functions from A to B . The sum in (ii) is countable because
of (i) and hence absolutely convergent. The function p can
be extended to the power set ℘Y by $p(Y') = \sum_{y \in Y'} p(y)$ yielding a σ-additive and normalized probability distribution.
Each function f':X → Y can be regarded as a (discrete)
stochastic channel f:X → Y defining f(x)(y) = 1   if
y = f'(x) and f(x)(y) = 0 otherwise.
In the literature d and l are often replaced by one
(discrete) stochastic channel t:SxI → SxO (cf. [2,7,20,75]).

<u>1.6 Definition</u> (B i l i n e a r   a n d   L i n e a r
A u t o m a t a) :   A <u>bilinear automaton</u> $A = (I,O,S,d,l)$
consists of R-modules I, O, S and R-bilinear functions
$d:S{\times}I \to S$ and $l:S{\times}I \to O$ where R is a commutative ring
with unit. Replacing d and l by R-linear functions we get
the notion of a <u>linear automaton</u>. Readers not familiar with
R-modules should consider only the special cases in which R
is a field or the ring of integers such that an R-module becomes a vector space or an abelian group respectively.

Using the universal property of the tensorproduct $A{\otimes}B$ of
R-modules each R-bilinear function $f:A{\times}B \to C$ can be reduced
to an R-linear function $\bar{f}:A{\otimes}B \to C$ in the following way:
For each pair of R-modules A,B there is a universal R-bilin-
ear function $u_{AB}:A{\times}B \to A{\otimes}B$ defined by $u_{AB}(a,b) = a{\otimes}b$ such
that for all R-modules C and R-bilinear functions $f:A{\times}B \to C$
there is a unique R-linear function $\bar{f}:A{\otimes}B \to C$ satisfying
$\bar{f}{\circ}u_{AB} = f$, i.e. diagram (1.6.1) is commutative.

(1.6.1)

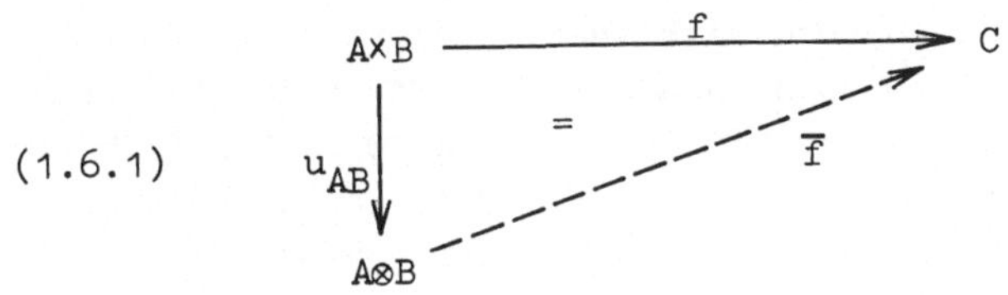

Hence the R-bilinear functions $d:S{\times}I \to S$ and $l:S{\times}I \to O$
given in the definition of bilinear automata can be regarded
as R-linear functions $\bar{d}:S{\otimes}I \to S$ and $\bar{l}:S{\otimes}I \to O$.
Henceforth we shall use this second notation for bilinear
automata.

In the case of linear automata each R-linear function
$d:S{\times}I \to S$ can be represented by two R-linear functions
$d_1:S \to S$ and $d_2:I \to S$ satisfying $d_1(s) = d(s,0)$ and
$d_2(x) = d(0,x)$ for $s{\in}S$ and $x{\in}I$. Vice versa we have $d(s,x) =$

$= d_1(s) + d_2(x)$. If I , 0 and S are finite dimensional
vector spaces the functions d_1, d_2 and similarly $l_1:S \to 0$,
$l_2:I \to 0$ representing $l:S \times I \to 0$ can be considered as
matrices.
Linear automata are discussed in [37,41,54,69] for example
and bilinear automata in [17,44].

1.7 <u>Definition</u> (T o p o l o g i c a l   A u t o m a t a) :
A <u>topological automaton</u> $A = (I,0,S,d,l)$ consists of topological spaces I , 0 , S and bicontinuous functions $d:S \times I \to S$
and $l:S \times I \to 0$.
In analogy to bilinear functions a bicontinuous function
$f:A \times B \to C$ is a function such that f is continuous in each
component separately, i.e. $f(a,-):B \to C$ and $f(-,b):A \to C$
are continuous for all $a \in A$, $b \in B$. Moreover each bicontinuous
function $f:A \times B \to C$ can be reduced to a continuous function
$\bar{f}:A \otimes B \to C$ using a universal property similar to (1.6.1).
In fact $A \otimes B$, called biproduct of A and B, is the cartesian
product of A and B endowed with the inductive or bitopology.
Explicitly a set 0 is open in $A \otimes B$ iff ("iff" means "if and
only if") for all $a \in A$ and $b \in B$ the sections $0_a = \{b \in B/(a,b) \in 0\}$
and $0_b = \{a \in A/(a,b) \in 0\}$ are open in B and A respectively.
Clearly each open set in the product topology of $A \times B$ is open
in $A \otimes B$, but vice versa the interior of the following cruci-
fix (1.7.1) together with the center is open in the bitopology but of course not in the product topology:

(1.7.1)

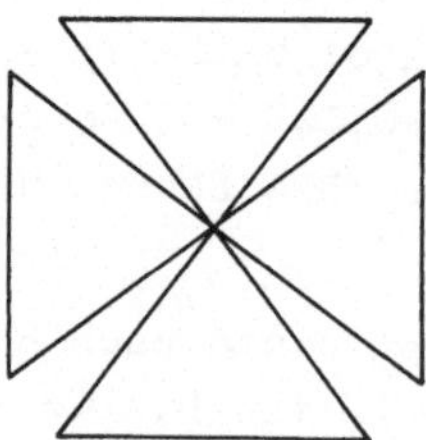

In the literature [14,15,67] topological automata are defined using continuous functions d:S×I → S and l:S×I → O
defined on the topological product of S and I. In fact our
notion which also is considered in [4] is more general and
in addition leads to better results. Only in the case of
locally compact input space I the results for both types of
products are similar.

1.8 Motivation: Now we want to give the proposed unified
representation for the different types of automata introduced up to now. In fact we only have to consider different
structures on the sets I, O, S and corresponding structure
preserving functions d and l . This at once leads to the
notion of a category K consisting of objects (e.g. sets,
R-modules or topological spaces) and morphisms (e.g. deterministic, R-linear or continuous functions). Moreover a category has an associative composition of morphisms - corresponding to the composition of functions - and for each object an identity morphism - corresponding to the identity
function - which is a unit element with respect to composition. For example we will have to consider the category of
sets with functions, of R-modules with R-linear functions
and the category of topological spaces with continuous functions.
The exact definition of a category (cf. 12.1) and other basic categorical notions are given in the appendix but motiva-
tions and examples will be given in the context.

Another difficulty is to give an appropriate generalization
of the cartesian product S×I, the tensorproduct and biproduct S⊗I in the R-module and the topological case respectively. Unfortunately we cannot take the categorical product
(cf. 12.4) because S⊗I fails to have the universal properties of a categorical product in the R-module and the topological case for example. The cartesian product is a categorical product with respect to deterministic functions but not
with respect to partial and nondeterministic functions.

This in fact is the background for the remark in 1.4. On the other hand in all our examples the cartesian, tensor and biproduct can be extended to structure preserving functions, i.e. for f:A → B , g:C → D we have f⊗g:A⊗C → B⊗D being compatible with composition and identities. Hence we have a bifunctor ⊗:$\underline{K}$x$\underline{K}$ → $\underline{K}$ in the sense of 12.5 where $\underline{K}$ is the category of sets, R-modules or topological spaces for example (cf. 1.10).

In fact, in all our examples we have a "monoidal category" which is a category $\underline{K}$ together with a bifunctor ⊗:$\underline{K}$x$\underline{K}$ → $\underline{K}$ in the following sense:

<u>1.9 Definition</u> (M o n o i d a l   C a t e g o r i e s) :
A <u>strict monoidal category</u> is a 3-tuple ($\underline{K}$,⊗,U) consisting of
- a category $\underline{K}$,
- a bifunctor ⊗:$\underline{K}$x$\underline{K}$ → $\underline{K}$, called tensor product,
- an object U of $\underline{K}$, called unit object,
such that we have for all objects A,B,C $\in$ $\underline{K}$:

(1.9.1)   A⊗U = A = U⊗A

(1.9.2)   A⊗(B⊗C) = (A⊗B)⊗C .

Replacing equality in (1.9.1) and (1.9.2) by natural isomorphisms r_A:A⊗U $\overset{\sim}{\to}$ A (right unit), l_A:U⊗A $\overset{\sim}{\to}$ A (left unit) and a_{ABC}:A⊗(B⊗C) $\overset{\sim}{\to}$ (A⊗B)⊗C (associativity) we get the notion of a <u>monoidal category</u> provided that the isomorphisms are coherent in the sense that the following diagrams (1.9.3), (1.9.4) and (1.9.5) are commutative for all objects A,B,C,D$\in$$\underline{K}$.

$$
\begin{array}{ccccc}
A⊗(B⊗(C⊗D)) & \overset{a}{\longrightarrow} & (A⊗B)⊗(C⊗D) & \overset{a}{\longrightarrow} & ((A⊗B)⊗C)⊗D \\
\downarrow{\scriptstyle A⊗a} & & = & & \uparrow{\scriptstyle a⊗D} \\
A⊗((B⊗C)⊗D) & & \overset{a}{\longrightarrow} & & (A⊗(B⊗C))⊗D
\end{array}
$$

(1.9.3)

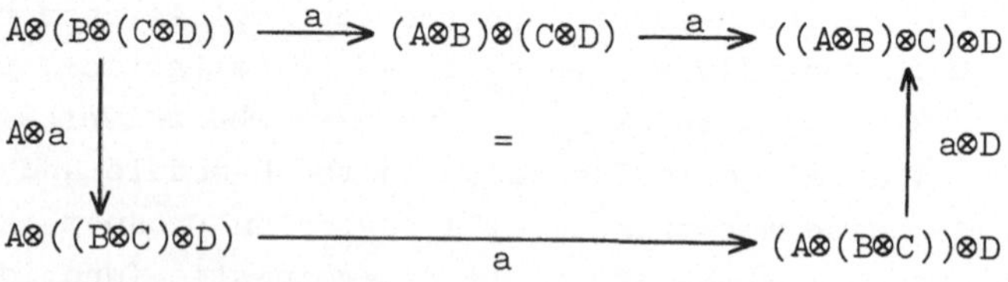

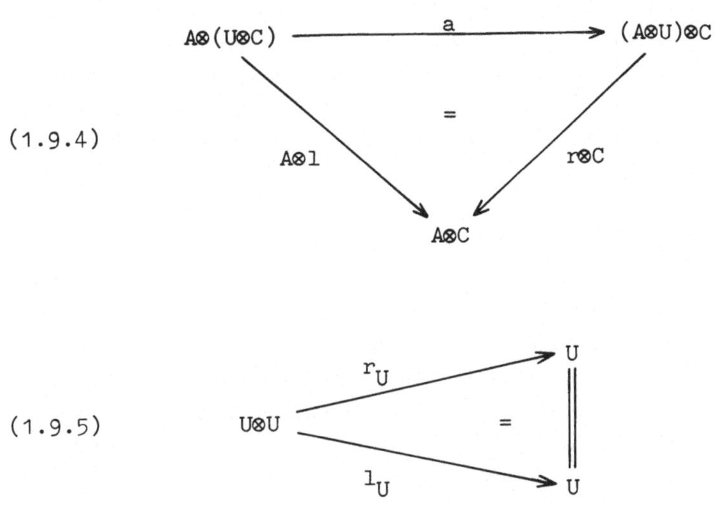

$$(1.9.4)$$

$$(1.9.5)$$

In (1.9.3) and (1.9.4) we have omitted the indices of a, l
and r and A⊗a is defined to be $id_A⊗a$ for example where id_A
is the identity morphism of the object A.

<u>Categorical Preliminaries</u>: Explicit definitions of catego-
ries, bifunctors, natural isomorphisms and commutative diagrams are given in the appendix 12.1 - 12.6.

<u>Remark</u>: Roughly speaking conditions (1.9.3), (1.9.4) and
(1.9.5) imply the commutativity of all diagrams built up by
r, l, a, ⊗ and identities only. For a detailed discussion
of these coherence properties the reader is refered to [59].

<u>General Convention</u>: Although all our examples of monoidal
categories are not strictly monoidal we will omit the isomorphisms r, l, a in most of our considerations. In fact
the main ideas become much more clear in this way and easier
to check especially for people in computer science. On the
other hand readers with more categorical background will

have no difficulties in filling in all the isomorphisms which are in fact coherent according to the above remark. Thus in our considerations we are dealing with strict monoidal categories but the theory is formulated and valid for arbitrary monoidal categories. Thus it is applicable to all of our examples. Instead of $(\underline{K},\otimes,U)$ we often will write $(\underline{K},\otimes)$ or only $\underline{K}$.

<u>1.10 Examples</u> (M o n o i d a l   C a t e g o r i e s) :

1. The category of sets consists of all sets as objects and all functions as morphisms. The composition of morphisms is exactly the composition of functions and the identity morphisms are the identity functions. The category of sets will be denoted by <u>Set</u> . Moreover the cartesian product of sets can be regarded as a bifunctor $\times:\underline{Set}\times\underline{Set}\to\underline{Set}$ because the cartesian product of functions satisfies the following equations (cf. 12.5):

(i) $(g\times g')\circ(f\times f')=(g\circ f)\times(g'\circ f'):A\times A'\to C\times C'$ for all
      $f:A\to B$, $g:B\to C$, $f':A'\to B'$, $g':B'\to C'$,

(ii) $id_A\times id_B = id_{A\times B}$ for arbitrary sets A and B.

Unfortunately the cartesian product of sets is not associative but only associative up to natural bijections. Moreover taking the one-element-set $U=\{1\}$ as unit object we only have natural bijections $r_A:A\times U \overset{\sim}{\to} A$, sending (a,1) to $a\in A$, and $l_A:U\times A \overset{\sim}{\to} A$ but not equality $A\times U = A = U\times A$. Hence $(\underline{Set},\times,U)$ cannot be strictly monoidal, but it is a monoidal category which is easily to be seen by checking conditions (1.9.3), (1.9.4) and (1.9.5) for arbitrary sets and elements.

2. The category $\underline{ND}$ of nondeterministic functions consists of all sets as objects and all nondeterministic functions as morphisms (cf. 1.4). The composition $g\circ f:A\to C$ of nondeterministic functions $f:A\to B$ and $g:B\to C$ is defined by

$$(g\circ f)(a) = \bigcup_{b\in f(a)} g(b) \quad \text{for all } a\in A,$$

i.e. the union of all sets g(b) ranging over all $b\in f(a)$.

It is easy to verify that this composition is associative
and coincides with the usual composition for deterministic
functions f and g. Identity morphisms are again the identity
functions. Thus Set is a subcategory of ND.
Moreover $(\underline{ND},\times,U)$ is a monoidal category taking $U = \{1\}$ and
regarding the cartesian product now as a bifunctor
$\times:\underline{ND}\times\underline{ND} \to \underline{ND}$ defined for nondeterministic functions $f:A \to B$
and $f':A' \to B'$ by $(f\times f')(a,a') = f(a)\times f'(a')$ for all $a\in A$,
$a'\in A'$.

3. The category Stoch of stochastic channels consists of
all sets as objects and (discrete) stochastic channels as
morphisms (cf. 1.5). Given stochastic channels $f:A \to B$,
$g:B \to C$ the composition $g\circ f:A \to C$ is a stochastic channel
defined for all $a\in A$, $c\in C$ by

$$(g\circ f)(a)(c) = \sum_{b\in B} [f(a)(b)]\cdot[g(b)(c)] .$$

Properties (i) and (ii) in 1.5 and associativity of the composition are easily verified using the absolute convergence
of the sums. According to the remark in 1.5 each function
can be regarded as a stochastic channel and especially the
identity functions are the identity morphisms in Stoch.
Moreover $(\underline{Stoch},\times,U)$ is a monoidal category taking again
$U = \{1\}$ and the cartesian product of sets, which can be extended to stochastic channels $f:A \to B$ and $f':A' \to B'$
defining for all $a\in A$, $b\in B$, $a'\in A'$, $b'\in B'$:

$$(f\times f')(a,a')(b,b') = f(a)(b)\cdot f'(a')(b') .$$

4. $\underline{Mod}_R$ is the category of all R-modules and R-linear functions (cf. 1.6) and $(\underline{Mod}_R,\otimes,R)$ is in fact a monoidal category where $\otimes$ is the usual tensor product of R-modules
regarded as a functor $\otimes:\underline{Mod}_R\times\underline{Mod}_R \to \underline{Mod}_R$. The tensor product $f\otimes f'$ of R-linear functions $f:A \to B$, $f':A' \to B'$ is
defined by:

$$(f\otimes f')(\sum_{i=1}^{n} r_i(a_i\otimes a_i')) = \sum_{i=1}^{n} r_i(f(a_i)\otimes f'(a_i'))$$

for $a_i\in A$, $a_i'\in A'$, $r_i\in R$ $(i = 1,\ldots,n)$.

On the other hand the following diagram (1.10.1) is commutative and in fact f⊗f' is uniquely determined by this property:

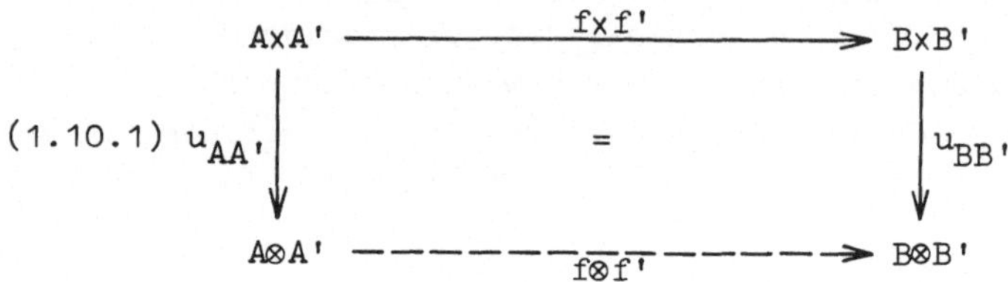

Moreover f⊗f' can be defined in the following way: Since $u_{BB'} \circ (f \times f')$ is R-bilinear f⊗f' in (1.10.1) is uniquely defined by the universal property of A⊗A' given in 1.6.
The bifunctor axioms can be checked more easily using this universal definition. Finally there are natural isomorphisms $r_A : A \otimes R \xrightarrow{\sim} A$, $l_A : R \otimes A \xrightarrow{\sim} A$ and $a_{ABC} : A \otimes (B \otimes C) \xrightarrow{\sim} (A \otimes B) \otimes C$ defined by $r_A(a \otimes r) = ra = l_A(r \otimes a)$ and $a_{ABC}(a \otimes (b \otimes c)) = (a \otimes b) \otimes c$ for a∈A, b∈B, c∈C on the generators. On the other hand conditions (1.9.3), (1.9.4) and (1.9.5) can be checked using the corresponding universal properties (cf. [59]). Let us remark that the isomorphism defined by sending a⊗(b⊗c) to -(a⊗b)⊗c would not satisfy condition (1.9.3).

<u>Remark</u>: Taking the direct product of R-modules we get another bifunctor $\times : \underline{Mod}_R \times \underline{Mod}_R \to \underline{Mod}_R$ so that $(\underline{Mod}_R, \times, U)$ becomes a monoidal category where U is now the one-element or null module.

5. The category <u>Top</u> of topological spaces consists of all topological spaces as objects and all continuous functions as morphisms (cf. 1.7). Using the same universal arguments as given in 4. for the tensor product it can be shown that $(\underline{Top}, \otimes, U)$ is a monoidal category where ⊗ is the biproduct of topological spaces and U the one-point space. Again there is another monoidal structure on <u>Top</u> taking the topological

product x of spaces which coincides with the categorical
product in <u>Top</u>. Hence we get another monoidal category
(<u>Top</u>,x,U).

<u>1.11 Definition</u> (A u t o m a t a   i n   M o n o i d a l
C a t e g o r i e s) : Given a monoidal category $(\underline{K},\otimes)$
an <u>automaton in</u> $(\underline{K},\otimes)$ is a 5-tuple $A = (I,O,S,d,1)$ where

   I, O and S are objects in <u>K</u> and

   $d:S\otimes I \to S$ and $1:S\otimes I \to O$ are <u>K</u>-morphisms.
The interpretation is the same as given in 1.1.

A <u>morphism of automata</u> $f:A \to A'$ is a triple $f = (f_I,f_O,f_S)$
of <u>K</u>-morphisms $f_I:I \to I'$, $f_O:O \to O'$ and $f_S:S \to S'$ satis-
fying $d'\circ(f_S\otimes f_I) = f_S\circ d$ and $1'\circ(f_S\otimes f_I) = f_O\circ 1$. Defining the
composition of automata morphisms in each component sepa-
rately we get the <u>category of automata</u> in $(\underline{K},\otimes)$.

<u>General Convention</u>: In the following chapters we will always
consider automata with fixed input and output objects I and
O. Hence automata will be written in the shorter form
$A = (S,d,1)$. A' for example in general denotes an automaton
with components $A' = (S',d',1')$ but again with fixed I and
O. Moreover it is natural to restrict the notion of automata morphisms $f:A \to A'$ to being <u>K</u>-morphisms $f:S \to S'$
satisfying $d'\circ(f\otimes I) = f\circ d$ and $1'\circ(f\otimes I) = 1$, i.e. $f_I = id_I$,
$f_O = id_O$ and $f_S = f$, in diagrams

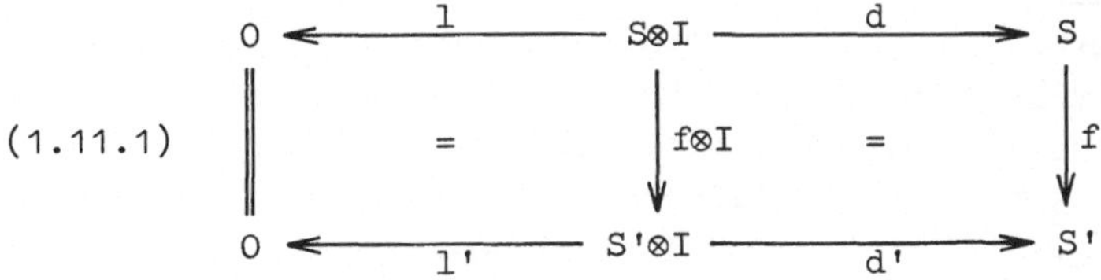

(1.11.1)

The composition of automata morphisms in this case is just
the composition in <u>K</u> which again leads to an automata mor-

phism (Note that we have $(f'\otimes I)\circ(f\otimes I) = (f'\circ f)\otimes I$ because of
the bifunctor properties of $\otimes$). The corresponding category
of automata in $(\underline{K},\otimes)$ with fixed I and O will be denoted by
$(\underline{K},\otimes)$-$\underline{\text{Aut}}$, or $\underline{K}$-$\underline{\text{Aut}}$ for short.

<u>Remarks</u>: 1. In the deterministic case our notion of automata
in 1.11 was first given by G. H. M e a l y . Thus it is
often called Mealy-type automaton. The notions of Medvedev-
and Moore-type automata will be given in 1.14.

2. Automata with fixed initial states given by a $\underline{K}$-morphism
$a:U \to S$ will be discussed in chapter 9. In most of our
examples U is a one-element-set $\{1\}$ and thus the initial
state $s_0 \in S$ is defined by $s_0 = a(1) \in S$ if a is a deterministic function for example.

3. In the case of nondeterministic and stochastic automata
for example it is useful to restrict the morphisms to be
deterministic functions, i.e. to belong to a subcategory
$\underline{K}'$ of $\underline{K}$. In this case the corresponding category of automata
will be written $\underline{K}$-$\underline{K}'$-$\underline{\text{Aut}}$.

<u>1.12 Unified Representation of Automata</u>: All the different
types of automata defined in 1.1 and 1.4 to 1.7 can be regarded as special cases of automata in monoidal categories.
Listing the examples given in 1.10 we have the following
monoidal categories $(\underline{K},\otimes)$:

$(\underline{\text{Set}},\times)$	for deterministic automata
$(\underline{\text{ND}},\times)$	for nondeterministic automata
$(\underline{\text{Stoch}},\times)$	for stochastic automata
$(\underline{\text{Mod}}_R,\times)$	for linear automata
$(\underline{\text{Mod}}_R,\otimes)$	for bilinear automata
$(\underline{\text{Top}},\otimes)$	for topological automata

Moreover the monoidal categories $(\underline{\text{PD}},\times)$, $(\underline{\text{Rel}},\times)$, $(\underline{\text{RelTop}},\otimes)$
of partial defined functions, relations and topological
relations, which will be defined in 1.13, lead to the following interesting types of automata:

($\underline{PD}$,×)	for partial automata
($\underline{Rel}$,×)	for relational automata
($\underline{RelTop}$,⊗)	for relational topological automata.

Sometimes it is useful to restrict the objects I, O and S as being finite sets, noetherian R-modules or compact topological spaces for example. This leads to the corresponding finite types of automata.

The above list can be split into two parts: deterministic type and nondeterministic type automata. For each part a common theory will be presented in the following chapters corresponding to the different power of the classical results which are valid for deterministic and nondeterministic automata respectively.

<u>Deterministic Type</u>: Deterministic, partial, linear, bilinear and topological automata.

<u>Nondeterministic Type</u>: Nondeterministic, stochastic, relational and relational topological automata.

These two types correspond to the notions of
<u>automata in closed</u> and <u>pseudoclosed monoidal categories</u>,
which will be studied in chapters 4, 5, 9 and 6, 7, 8, 9, 10 respectively. Moreover several other examples will be mentioned at the end of chapters 4 and 6 for the deterministic and the nondeterministic type separately.

<u>1.13 Examples</u>: 1. (P a r t i a l   A u t o m a t a) :
The category $\underline{PD}$ of partially defined functions, short partial functions, consists of sets as objects and partial functions as morphisms. A partial function f:A → B is a deterministic function defined on a subset A' of A only and undefined otherwise. The composition of partial functions f:A → B and g:B → C is a partial function g∘f:A → C which is exactly defined for those a∈A for which f(a) and g(f(a)) are defined. Together with the cartesian product we get a monoidal category ($\underline{PD}$,×) and automata in ($\underline{PD}$,×) are exactly partial automata.

2. (R e l a t i o n a l   A u t o m a t a) :   A relation f
on A×B is a subset f⊆A×B and will be denoted by f:A → B.
Thus it can be regarded as a partially defined, nondeterministic function f from A to B, or as a function f' from
A to the powerset ℘B of B defined by f'(a) = {b∈B/ afb } for
all a∈A. Note that afb as usual means (a,b)∈f. The well
known composition of relations f:A → B, g:B → C leads to a
relation g∘f:A → C defined by: a(g∘f)c iff there is a b∈B
satisfying afb and bgc. This composition coincides with
that of nondeterministic, partially defined and deterministic
functions in the corresponding special cases. Hence we get
the category _Rel_ of relations with sets as objects and relations as morphisms. Moreover (_Rel_,×) is a monoidal category
where × is the cartesian product defined for relations
componentwise. Of course automata in (_Rel_,×) are the well
known relational automata (cf. [12,13,70]).

3. (R e l a t i o n a l   T o p o l o g i c a l   A u t o -
m a t a) :   There are several notions of continuity for relations known in the literature but not all of them are
appropriate with respect to automata theory (cf. [14,15,32]).
We want to consider only lower semicontinuous relations
meaning that the inverse image of open sets is open. Note
that this condition does not imply that the inverse image of
closed sets is closed as it is well known in the case of
functions. Now the category _RelTop_ has as objects topological
spaces and as morphisms lower semicontinuous relations.
Moreover (_RelTop_,⊗) is a monoidal category taking the biproduct ⊗ of topological spaces defined in 1.7. The biproduct
of relations is just the cartesian product which again leads
to a lower semicontinuous relation f⊗g if the relations f
and g have this property. Automata in (_RelTop_,⊗) will be
called relational topological automata. For more details
concerning other types of relational topological automata we
refer the reader to [14,15,32] and to the end of chapter 6.

<u>1.14 Medvedev- and Moore-Automata</u>: Finally we will give the notion of a Medvedev-automaton which is just the state-transition part of an automaton in the sense of 1.11 and we will make a remark concerning Moore-automata:

<u>Definition</u> (M e d v e d e v - A u t o m a t a) :
A <u>Medvedev-automaton</u> in a monoidal category $(\underline{K}, \otimes)$ is a triple $A = (I, S, d)$ consisting of the objects I and S and a $\underline{K}$-morphism $d: S \otimes I \to S$. A morphism $f: A \to A'$ of Medvedev-automata is a pair $f = (f_I, f_S)$ satisfying $d' \circ (f_S \otimes f_I) = f_S \circ d$.

<u>General Convention</u>: In most cases we will fix the input object I so that Medvedev-automata are denoted by $A = (S, d)$ and morphisms by $f = f_S$ only. The corresponding category of Medvedev-automata (with fixed I) is written $(\underline{K}, \otimes)$-<u>Medv</u> or $\underline{K}$-<u>Medv</u> for short.

<u>Remark</u> (M o o r e - A u t o m a t a) : In all our considerations it is possible to replace Mealy-automata, defined in 1.11, by Moore-automata where $l: S \otimes I \to O$ is replaced by a morphism $m: S \to O$ so that the output is independent of the current input. But it is left to the reader to reformulate the theory for Moore-automata.

2. Some Problems in Automata Theory

In this chapter we want to formulate the problems of equivalence, reduction, minimization and realization as well as some basic notions concerning the transition monoid and the structure theory of automata. In order to give a motivation for the constructions in the following chapters we will sketch the problems and the corresponding results for the case of deterministic automata in such a way that they can be generalized to automata in monoidal categories of deterministic type (cf. 1.12). For a categorical theory of decomposition and synthesis of automata we refer the reader to [18]. Unless otherwise specified automata are always deterministic in this chapter.

<u>2.1 Input-Output Behavior and Equivalence of Automata:</u>
Given an automaton $A = (S,d,l)$ we start with a universal construction of the extended output function $l^+ : S \times I^+ \to 0$ (cf. 1.2). For all natural numbers $n \in \mathbb{N}$ $l_n : S \times I^n \to 0$ is defined recursively by

$$l_1 = l : S \times I \to 0 \qquad \text{and}$$

$$l_{n+1} = (S \times I^{n+1} \xrightarrow{\ dx I^n\ } S \times I^n \xrightarrow{\ l_n\ } 0).$$

Hence there is a unique function $\bar{I} : S \times (\underset{n \in \mathbb{N}}{\overset{\cdot}{\cup}} I^n) \to 0$ defined on the disjoint union such that $\bar{I}$, restricted to $S \times I^n$, is equal to l_n. Moreover we have the isomorphism

$$\underset{n \in \mathbb{N}}{\overset{\cdot}{\cup}} I^n \; \overset{\sim}{=} \; I^+$$

by definition of the disjoint union ranging over all natural numbers $n \in \mathbb{N}$. Hence l^+, defined by

$$l^+ = (S \times I^+ \overset{\sim}{\to} S \times (\underset{n \in \mathbb{N}}{\overset{\cdot}{\cup}} I^n) \overset{\bar{I}}{\to} 0),$$

coincides with l^+ defined in 1.2.

Defining $<I^+,O>$ to be the set of all functions from I^+ to
O we get a new function

$$M(A):S \longrightarrow <I^+,O>$$

defined by $M(A)(s)(w) = 1^+(s,w)$ for $s\in S$, $w\in I^+$, called
__machine function__ M(A) of A. Clearly M(A) assigns to each
state $s\in S$ the corresponding input-output behavior
$M(A)(s) = 1^+(s,-):I^+ \to O$. The set of all these input-output
functions is called input-output behavior of A .

__Definition__: Let $M(A):S \to <I^+,O>$ be the machine function
of an automaton A then the image E(A) of M(A) , i.e.

$$E(A) = \{M(A)(s):I^+ \to O/\ s\in S\} \subseteq <I^+,O> ,$$

is called __input-output behavior__ of A, or short __behavior__ of
A. Automata A and A' are called __equivalent__ if they have
the same behavior, i.e. $E(A) = E(A')$.

__Remark__: There is not only a universal construction for
$1^+:SxI^+ \to O$ but also for the machine function M(A). Given
1^+ and the evaluation function $ev:<I^+,O>xI^+ \to O$, defined
by $ev(f,w) = f(w)$ for $f:I^+ \to O$ and $w\in I^+$, M(A) is the
unique function $M(A):S \to <I^+,O>$ such that the following
diagram is commutative:

(2.1.1)

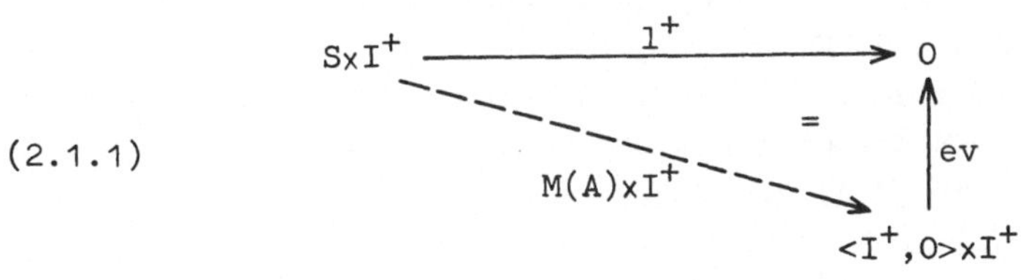

In fact the commutativity is equivalent to $1^+(s,w) =$
$= ev \circ (M(A)xI^+)(s,w) = ev(M(A)(s),w) = M(A)(s)(w)$ which is
just the definition of M(A). Such a universal construction
of 1^+ and M(A) will be used later on for automata in
monoidal categories.

<u>2.2 Reduction and Minimization</u>: Given an automaton A we
now are going to consider the problem of constructing an
automaton A' equivalent to A with minimal number of
states. There are in fact two different methods of doing
this:

<u>Reduction</u>: States $s,s' \in S$ are called equivalent, written
$s \equiv s'$, if they have the same input-output behavior, i.e.
$M(A)(s) = M(A)(s')$. Equivalence of states defines an equivalence relation $\equiv$ on S, called Nerode-equivalence.

<u>Problem 1</u>: Is it possible to factorize the set of states S
by the Nerode-equivalence in order to get an equivalent
automaton A' with state object $S_{/\equiv}$ which is a homomorphic image of A ?

<u>Minimization</u>: By definition the cardinality of the behavior
$E(A)$ is less than or equal to the cardinality of the states
of A .

<u>Problem 2</u>: Is there an automaton A' equivalent to A with
state set equal or isomorphic to $E(A)$?

<u>Solution of the problems</u>: Factorizing S by the Nerode-
equivalence, which is the equivalence relation caused by the
machine function $M(A)$, we get a surjective function
$e : S \to S_{/\equiv}$ and an injective supplement $m : S_{/\equiv} \to \langle I^+, O \rangle$ satisfying $m \circ e = M(A)$. On the other hand we have the image fac-
torization of $M(A)$ given by $e(A) : S \to E(A)$ and the in-
clusion $m(A) : E(A) \to \langle I^+, O \rangle$, i.e. $m(A) \circ e(A) = M(A)$. Since
we have a unique image factorization up to isomorphism there
is a unique bijection of such a kind that the following diagram (2.2.1) is commutative. Hence $\bar{S} := S_{/\equiv}$ and $E(A)$ are
isomorphic and problems 1 and 2 are solved by the following theorem:

<u>Theorem</u>: Given an automaton A and a factorization
$S \xrightarrow{e} \bar{S} \xrightarrow{m} \langle I^+, O \rangle$ of the machine function $M(A)$ as
above there are unique functions $\bar{d} : \bar{S} \times I \to \bar{S}$ and $\bar{I} : \bar{S} \times I \to O$

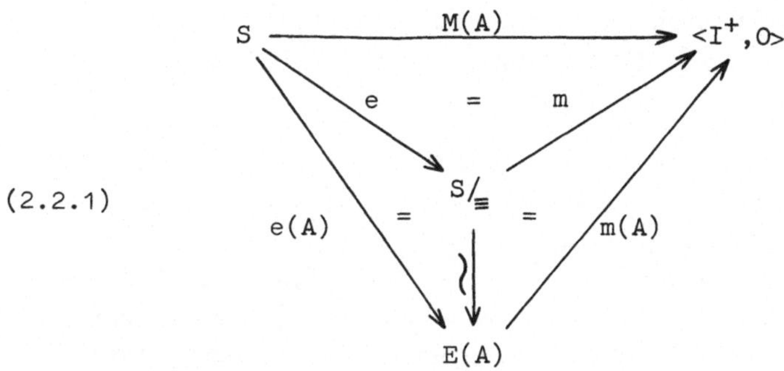

(2.2.1)

so that $\bar{A} = (\bar{S}, \bar{d}, \bar{I})$ is an automaton with injective machine
function $M(\bar{A})$ equal to m and $e: A \to \bar{A}$ is a reduction,
i.e. a surjective automata morphism. Hence $\bar{A}$ is equivalent
to A and solves problems 1 and 2 .

In order to prove the theorem we use the following properties of the machine function $M(A)$:

<u>2.3 Lemma</u> (M a c h i n e   F u n c t i o n s) :
The machine function $M(A)$ of an automaton A is characterized as a function $g: S \to <I^+, O>$ such that the following
diagrams (2.3.1) and (2.3.2) are commutative:

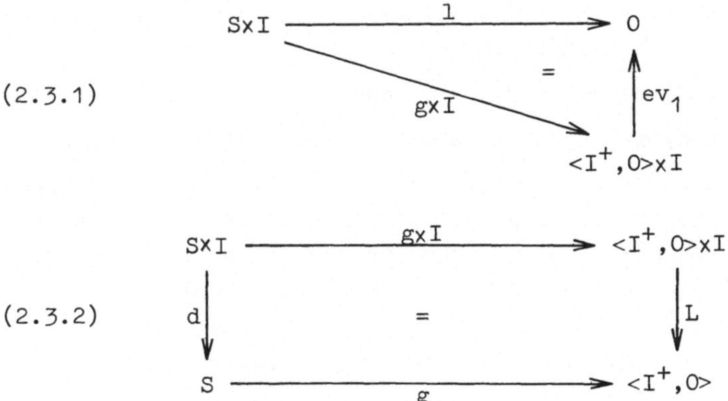

(2.3.1)

(2.3.2)

where the evaluation ev_1 and the left shift L are defined by $ev_1(f,x) = f(x)$ resp. $L(f,x) = f \circ L_x$ with
$L_x(w) = xw$ for all $f \in \langle I^+, O \rangle$ and $x \in I$.

Note that (2.3.2) is just the condition for g being a
Medvedev-automata morphism (cf. 1.14). Moreover we have the
following diagonal condition:
Given a factorization $S \xrightarrow{e} \bar{S} \xrightarrow{m} \langle I^+, O \rangle$ of M(A) with
surjective e and injective m there is a unique diagonal
function $\bar{d} : \bar{S} \times I \to \bar{S}$ making diagram (2.3.3) commutative:

(2.3.3)

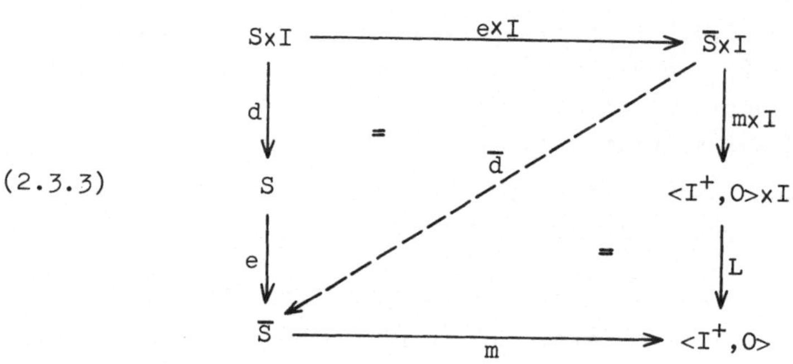

Proof: Commutativity of the diagrams (2.3.1) and (2.3.2) is
equivalent to the conditions

(2.3.4) $g(s)(x) = 1(s,x)$ and

(2.3.5) $g(s)(xw) = g(s) \circ L_x(w) = L \circ (g \times I)(s,x)(w) = g(d(s,x))(w)$

for all $s \in S$, $x \in I$, $w \in I^+$. Moreover the function g is uniquely defined by (2.3.4) and (2.3.5) which is in fact a recur-
sive definition for g . On the other hand the machine function M(A) defined by $M(A)(s)(w) = 1^+(s,w)$ satisfies
(2.3.4) and (2.3.5) because we have for $w \in I^+$ of length n
$$M(A)(s)(xw) = 1^+(s,xw) = 1_{n+1}(s,xw) = 1_n(d(s,x),w) =$$
$$= 1^+(d(s,x),w) = M(A)(d(s,x))(w) \; .$$

Hence $M(A)$ also satisfies (2.3.1) and (2.3.2) and is characterized by this property.

In order to prove the diagonal condition let us first notice
that the outer diagram in (2.3.3) is commutative by (2.3.2)
and $m \circ e = M(A)$. Since e is surjective and m is injective
there is a unique function $\bar{d}: \bar{S} \times I \to \bar{S}$ making (2.3.3) commutative. Define $\bar{d}(\bar{s}, x) = e \circ d(s, x)$ for $x \in I$ and arbitrary
$s \in S$ satisfying $e(s) = \bar{s}$. It is easy to show that $\bar{d}$ is
well-defined and unique using the injectivity of m . ∎

<u>Proof of the Theorem in 2.2</u>: Given $\bar{d}$ and $\bar{I}$ such that
$M(\bar{A}) = m$ and $e: A \to \bar{A}$ is an automata morphism, $\bar{d}$ is the
unique supplement in diagram (2.3.3) by the diagonal condi-
tion using diagram (2.3.2) for $\bar{A}$. Moreover $\bar{I}$ is equal to
$ev \circ (M(\bar{A}) \times I) = ev \circ (m \times I)$ by (2.3.1). Vice versa defining $\bar{d}$ by
(2.3.3) and $\bar{I} = ev \circ (m \times I)$ m satisfies (2.3.1) and (2.3.2)
with respect to $\bar{A}$ and thus it is equal to $M(\bar{A})$. Moreover
e is an automata morphism because of the left triangle in
(2.3.3) and $\bar{I} \circ (e \times I)(s, x) = \bar{I}(e(s), x) = M(\bar{A})(e(s))(x) =$
$= m(e(s))(x) = m \circ e(s)(x) = M(A)(s)(x) = 1(s, x)$ for all $s \in S, x \in I$.
Finally $M(\bar{A}) = m$ and $M(A) = m \circ e$ implies $E(A) = E(\bar{A})$ and
therefore $\bar{A}$ solves problems 1 and 2 in 2.2. ∎

The above constructed $\bar{A}$ does not only have minimal number
of states but is also observable and reduced in the following
sense.

<u>2.4 Observable and Reduced Automata</u>: An automaton A is
called <u>observable</u> if the machine function $M(A)$ is injective. A is called <u>reduced</u> if for arbitrary A' each surjective automata morphism $f: A \to A'$, which will be called
<u>reduction</u>, is already an isomorphism, i.e. f is bijective.

<u>Remark</u>: Clearly a finite automaton A has minimal number of
states (equal to the cardinality of $E(A)$) iff A is observable (cf. 2.2, 3.12). This does not remain true for infinite automata where minimal cardinality of states is a much

weaker condition. But in any case we have the following
characterization:

Theorem: Given an automaton A' the following conditions
are equivalent:

(i)   A' is observable

(ii) A' is reduced

(iii) For all automata A satisfying $E(A) \subseteq E(A')$ there
      is a unique automata morphism $f:A \to A'$ which is
      surjective in the case $E(A) = E(A')$.

Moreover the reduction $e:A \to \bar{A}$ given in the theorem in 2.2
has the following universal property:

(iv) For all reduced automata A' and all automata mor-
      phisms $f:A \to A'$ there is a unique automata mor-
      phism $\bar{f}:\bar{A} \to A'$ satisfying $\bar{f} \circ e = f$.

Proof: The proof is a corollary of the theorem in 2.2 and
lemma 2.3 using in addition the fact that $f:S \to S'$ is an
automata morphism $f:A \to A'$ iff f satisfies $d' \circ (f \times I) = f \circ d$
and $M(A) = M(A') \circ f$. But we do not give the details because
this theorem will be a special case of results for automata
in closed categories which will be given in chapter 5
(cf. 5.5).

2.5 Behavior Characterization and Realization: The behavior
of an automaton A is a subset E(A) of $\langle I^+, O \rangle$ by definition. The problem is now to characterize those subsets B
of $\langle I^+, O \rangle$ which can be realized by an automaton A, i.e.
$E(A) = B$. The solution is a direct consequence of 2.3:

Theorem: A subset B of $\langle I^+, O \rangle$ is realizable by an auto-
maton iff B is closed under left shift $L: \langle I^+, O \rangle \times I \to \langle I^+, O \rangle$,
i.e. for all $f \in B$, $x \in I$ we have $L(f,x) = f \circ L_x \in B$.

Proof: The above condition is equivalent to the existence of
a function $L_B: B \times I \to B$ satisfying $L \circ (m \times I) = m \circ L_B$ where

$m:B \rightarrow \langle I^+,O \rangle$ is the inclusion. Thus given an automaton A
we have this condition for $E(A) = \bar{S}$ in (2.3.3) of lemma 2.3.
Vice versa given such a B it follows from (2.3.1) and
(2.3.2) that m is the machine function of the automaton
$\bar{A} = (B,L_B,l_B)$ with $l_B = ev_1 \circ (m \times I)$ and hence $E(\bar{A}) = B.$ ∎

2.6 <u>Further Problems Concerning Reduction, Minimization,
Equivalence and Realization</u>: The following list of problems
which are not only interesting for automata in our sense but
also for other types of automata, machines and systems will
be studied in chapter 3 in a much more general framework.
For deterministic automata some of these problems have already been solved before or they are direct consequences of
previous results.

1. General definition and characterization of reduced
   automata and uniqueness of reduced automata

2. Decomposition of the reduction process

3. General definition, characterization and uniqueness of
   minimal automata

4. Characterization of equivalent automata

5. Construction, characterization and uniqueness of minimal and finite minimal realizations.

In fact we do not always have such a nice situation where
reduction and minimization coincide up to isomorphism as it
is in the case of deterministic automata (cf. 2.2, 2.4).
We will see in the following examples of nondeterministic
automata in 2.7 that reduction and minimization do not lead
to the same results in general. Moreover the construction of
an equivalent observable automaton is not unique up to isomorphism.

2.7 <u>Reduction and Minimization of Nondeterministic Automata</u>:
Given a nondeterministic automaton $A = (S,d,l)$ the extended
output function $l^+:S \times I^+ \rightarrow O$ can be constructed in the same

way as given in 2.1 now for the case of nondeterministic
functions. Due to 1.4 the nondeterministic function 1^+ corresponds to a deterministic function $S \times I^+ \to P'O$ yielding a
deterministic machine function $M(A):S \to \langle I^+, P'O \rangle$ in the
same way as given in 2.1 for deterministic automata. Moreover the behavior $E(A)$ of A again is the image of the
machine function $M(A)$. But unfortunately the solution of
the reduction and minimization problem given in 2.2 cannot
be applied in the nondeterministic case because the central
lemma 2.3 is no more valid. But there is an uncanonical way
of constructing a state transition function $\bar{d}: \bar{S} \times I \to \bar{S}$ with
$\bar{S} := E(A) \cong S/_\equiv$ (cf. 2.2.1) using a coretraction of the surjective function $e(A):S \to \bar{S}$ which exists by the axiom of
choice. A coretraction is an arbitrary function $c: \bar{S} \to S$
satisfying $e(A) \circ c = id_{\bar{S}}$ which of course is not uniquely determined. Now we can define

$$(2.7.1) \qquad \bar{d} := e(A) \circ d \circ (c \times I) \quad \text{and} \quad \bar{I} := 1 \circ (c \times I)$$

and in fact we get an equivalent nondeterministic automaton
$\bar{A} = (\bar{S}, \bar{d}, \bar{I})$ which is observable in the sense of 2.4. Due to
the arbitrary choice of the coretraction $\bar{A}$ is not unique
up to isomorphism and $e(A):S \to \bar{S}$ does not define an automata morphism. On the other hand there is a different construction yielding an equivalent reduced automaton $R(A)$ in
the sense of 2.4 which is in fact a homomorphic image of A.
We state these results without proofs for the moment but
they will be given in chapter 7 in the more general framework of automata in pseudoclosed categories.

<u>Theorem</u>: For each nondeterministic automaton A there is
   (i) an equivalent observable automaton A' with state
       set isomorphic to the behavior $E(A)$

   (ii) an equivalent reduced automaton $R(A)$ and a reduc-
       tion $u(A):A \to R(A)$.

Moreover $u(A)$ satisfies the universal property given in
(iv) of 2.4 but $R(A)$ in general is not observable.

<u>Examples</u>: Consider again the nondeterministic automaton A_2
defined in (1.3.2) and in (2.7.2) respectively:

(2.7.2)

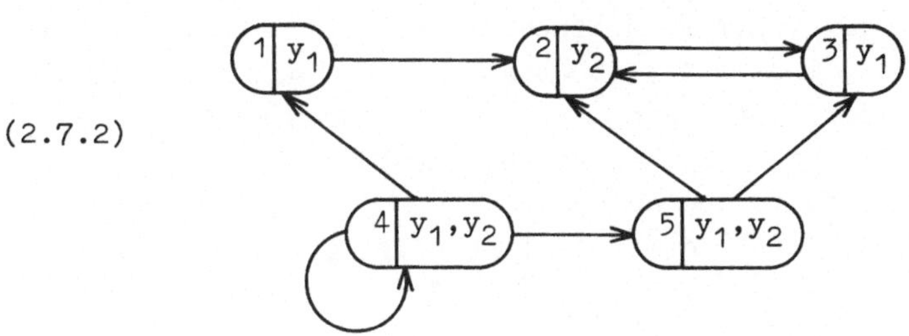

For $1_2^+ : S_2 \times I^+ \to 0$ we have $1_2^+(1,x^n) = 1_2^+(2,x^{n+1}) = 1_2^+(3,x^n) =$

$= \begin{cases} \{y_1\} & \text{if n is odd} \\ \{y_2\} & \text{if n is even} \end{cases}$ and $1_2^+(4,x^n) = 1_2^+(5,x^n) = \{y_1,y_2\}$.

Thus the states 1 and 3 as well as 4 and 5 are equivalent respectively and the behavior $E(A_2)$ contains the
three elements $b = 1_2^+(1,-) = 1_2^+(3,-)$, $b' = 1_2^+(2,-)$ and
$b'' = 1_2^+(4,-) = 1_2^+(5,-)$. Identifying the states 1 and 3 we
get the automaton A_3 given by (2.7.3):

(2.7.3)

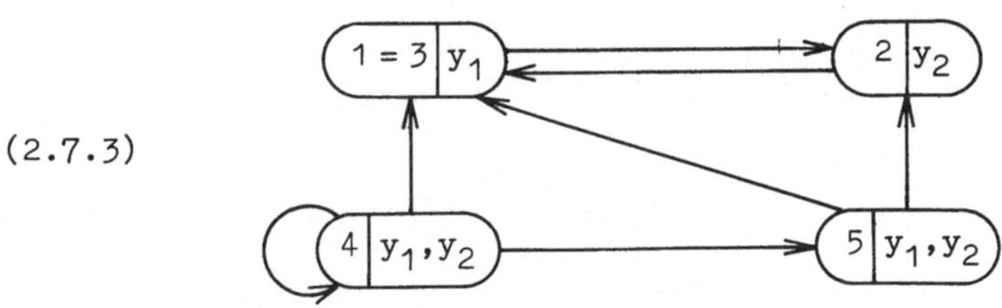

The identification function is a surjective automata morphism and hence a reduction. Obviously there is no reduction
which identifies the equivalent states 4 and 5 because
they have different state transitions with respect to equivalence of states. Hence A_3 is the reduced automaton of A_2 ,

but A_3 is not observable.

In order to construct observable automata which are equivalent to A_2 we study coretractions of the surjective function $e(A_2):S_2 \to E(A_2)$ defined by $e(A_2)(1) = e(A_2)(3) = b$, $e(A_2)(2) = b'$ and $e(A_2)(4) = e(A_2)(5) = b''$ according to the above consideration. Using (2.7.1) the coretraction $c:E(A_2) \to S_2$ of $e(A_2)$ given by $c(b) = 1$, $c(b') = 2$ and $c(b'') = 4$ leads to the automaton $A_4 = (I,O,E(A_2),e(A_2) \circ d_2 \circ (c \times I),l_2 \circ (c \times I))$ given by (2.7.4):

(2.7.4)

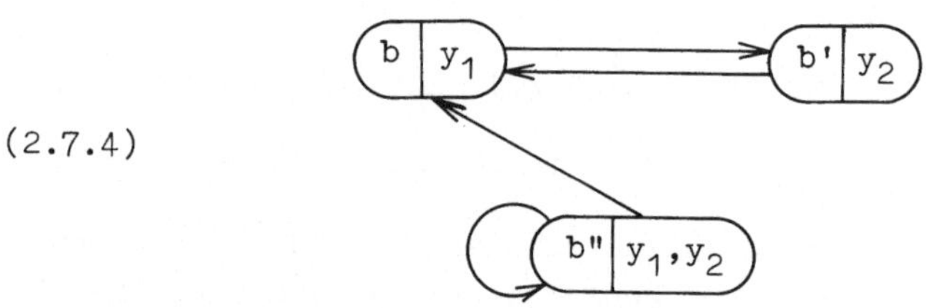

Choosing $c'(b) = 1$, $c'(b') = 2$ and $c'(b'') = 5$ we get another coretraction c' of $e(A_2)$ and the automaton A_5 in (2.7.5):

(2.7.5)

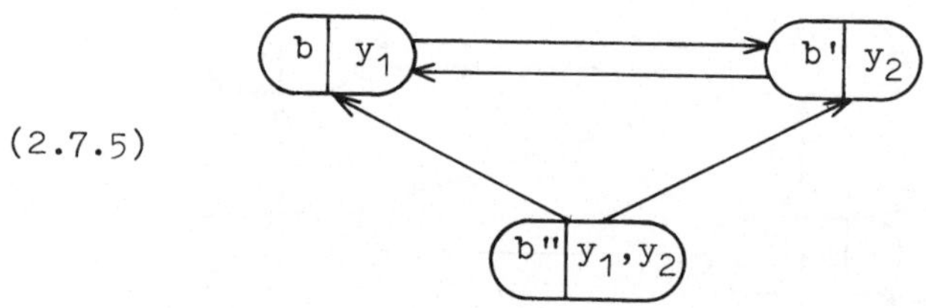

Obviously A_4 and A_5 are not isomorphic but both of them are observable and equivalent to A_2 in (2.7.2).

Now we turn back to consider deterministic automata again.

2.8 <u>Transition Monoids of Automata</u>: Given an automaton
$A = (S,d,l)$ each input string $w \in I^*$ yields a transition
$d_w : S \to S$ of states defined by $d_w(s) = d^*(s,w)$ for all $s \in S$.
The set of all these transitions $T(A) = \{d_w : S \to S / w \in I^*\}$ is
called the <u>transition monoid</u> of A. In fact, taking the usual
composition of functions, $T(A)$ is a monoid with unit
$d_\square = id_S$. Moreover $\cdot T(A)$ can be shown to be a factor monoid
of the free monoid I^*.

Of course, concerning the transition monoid only it suffices
to regard Medvedev-automata (cf. 1.14) because the output
function is not needed. Moreover it makes sense to replace
the input object I by an arbitrary monoid M such that
the state transition function $d : S \times M \to S$ for all $s \in S$ and
$w', w \in M$ with multiplication $w' \cdot w$ satisfies

$$d(s,1) = s \qquad (1 \in M \quad \text{unit})$$

$$d(s, w' \cdot w) = d(d(s,w'), w).$$

Clearly, taking the free monoid I^*, these axioms are satisfied for $d^* : S \times I^* \to S$ (cf. 1.2).

<u>Definition</u>: A <u>monoid automaton</u> is a triple $A = (M,S,d)$
where M is a monoid (with multiplication $m : M \times M \to M$ and
unit $1 : \{1\} \to M$), S is a set and $d : S \times M \to S$ is a function
satisfying

$$d \circ (S \times 1) = id_S \quad \text{and}$$

$$d \circ (S \times m) = d \circ (d \times M).$$

Now $d : S \times M \to S$ leads to the <u>transition function</u>
$t(A) : M \to \langle S, S \rangle$ defined by $t(A)(w)(s) = d(s,w)$. The image
$T(A)$ of the transition function $t(A)$ is called <u>transition</u>
<u>monoid</u> of A which is in fact a submonoid of $\langle S, S \rangle$ being
the monoid of endofunctions on S .
In the case $M = I^*$ both notions of transition monoids coincide. In fact the transition monoid can be regarded as another sort of behavior of an automaton and hence it is natural to study the problems of reduction, minimization, equivalence and realization again (cf. 2.6). In fact there is up

to isomorphism a unique minimal and reduced monoid automaton $\bar{A} = (T(A),S,\bar{d})$ equivalent to A , meaning now $T(\bar{A}) = T(A)$, where $\bar{d}:T(A)\times S \to S$ is the restriction of the evaluation $ev:<S,S>\times S \to S$. These problems will be studied in chapter 5 (cf. 5.6 and 5.7).

<u>2.9 Structure Theory of Automata</u>: Similar to groups or monoids etc., automata can be regarded as algebras in the sense of universal algebra. Automata of course are hetero-geneous algebras because there are three basic sets I, O and S. The operations are $d:S\times I \to S$ and $l:S\times I \to O$ and there are no axioms. Now in category theory structural problems have already been studied in great generality. So it suffices to prove whether certain categorical constructions exist in the category of automata and to characterize them in terms of automata theory. For deterministic automata this already has been done in [33] in great detail. In the list below corresponding notions of group and automata theory are given in brackets but for the categorical definitions we refer to the appendix, chapter 12. These constructions and corresponding characterizations for automata in monoidal categories are given in chapter 11.

1. Isomorphisms (Isomorphisms of groups resp. automata)

2. Subobjects   (Subgroups resp. subautomata)

3. Equalizers   (Kernels of homomorphisms)

4. Kernel pairs (Normal subgroups resp. automata con-
                 gruences)

5. Inverse images (- of groups resp. automata)

6. Products   (Products of groups resp. parallel com-
                 position of automata)

7. Coequalizers (Quotient groups resp. automata)

8. Coproducts   (Free products of groups resp. union of
                 automata)

9. Image factorizations (- of groups resp. automata)

10. Free constructions   (Free groups resp. free automata).

3. General Concepts of Reduction, Minimization and Realization

In this chapter we want to find a general setting for notions like "behavior", "minimal" and "reduced" in order to
formulate some general concepts of reduction, minimization
and realization and especially their relationships. All
these notions were introduced in the last chapter, concerning the input-output behavior as well as the transition
monoid of automata. Moreover in the literature these notions
have also been used for several other kinds of automata,
machines and systems unfortunately mostly with different
meaning. Motivated by the idea of minimal realization given
in [45], which will also be extended to cover the nondeterministic cases, we are going to study the concepts of reduction, minimization and realization in the general setting of
categories and functors. The explicit constructions for
automata of deterministic and nondeterministic type listed
in 1.12 will be given in the following chapters. In fact
this chapter is a short version of [30] where in addition
several other examples (cf. [3,5,8,35,36,45,53,56,72]) were
studied and classified with respect to this general concept.

In view of our purpose it suffices to regard a class of automata or systems as a class of objects in a category with
automata or system morphisms corresponding to the morphisms
in a category. Furthermore, the behavior construction can be
regarded as a functor from the category of systems to the
partially ordered class of all behaviors, called behavior
category. In fact each partially ordered class $(S,\leq)$ actually is a category with objects corresponding to the elements
$a,b\in S$ and with exactly one morphism from a to b , iff we
have $a\leq b$ and no morphism otherwise. Let us recall that a
functor $F:\underline{K} \to \underline{L}$ between the categories $\underline{K}$ and $\underline{L}$ defines

for each object A in $\underline{K}$ an object F(A) in $\underline{L}$ and for
each morphism f:A → B in $\underline{K}$ a morphism F(f):F(A) → F(B)
in $\underline{L}$ such that F preserves composition and identities
(cf. 12.5).

$\underline{3.1 \ Concept \ of \ Systematics}$: In the case of deterministic
automata we are concerned with the category $\underline{Aut}$ of deterministic automata (cf. 1.11 with $(\underline{K},\otimes) = (\underline{Set},x)$) and the
behavior category $\underline{B}$ of all behaviors E(A) which are
subsets of $<I^{+},O>$ and partially ordered by inclusion
(cf. 2.1). Now let us remark that each morphism f:A → A'
of automata implies an inclusion $E(A) \subseteq E(A')$ of the corresponding behaviors, which will be shown in the next chap-
ter, such that $E:\underline{Aut} → \underline{B}$ can be regarded as a functor.
The functor axioms F1 and F2 are obvious in the case of
a partially ordered class $\underline{B}$. In a similar way, using
remark 3 in 1.11, the behavior construction in the case of
nondeterministic automata can be shown to be a functor
$E:\underline{ND}-\underline{Set}-\underline{Aut} → \underline{B}$ where $\underline{B}$ is now the partially ordered set
of all nondeterministic behaviors which are subsets of
$<I^{+},P'O>$ (cf. 2.7). Finally we will show in 5.7 that the
construction of the transition monoid given in 2.8 can be
regarded as a functor from the category of monoid automata
to a corresponding behavior category consisting of subsets
of $<S,S>$.

Now a collection of systems together with a fixed behavior
construction will be called a "systematic", because systems
can be systematized by their behaviors. For example systems
are "equivalent" or "comparable" if their behaviors are
equal or comparable with respect to the partial ordering $\subseteq$
of the behavior objects.

$\underline{Definition}$ (S y s t e m a t i c) : A $\underline{systematic}$ is a
triple $\underline{S} = (\underline{S},\underline{B},E)$ consisting of a category $\underline{S}$, called
$\underline{system \ category}$, a partially ordered class $\underline{B}$, called $\underline{behavior}$
$\underline{category}$, and a functor $E:\underline{S} → \underline{B}$, called $\underline{behavior \ functor}$.

A <u>subsystematic</u> $\underline{S}' = (\underline{S}', \underline{B}, E')$ of $\underline{S}$ is a systematic such that $\underline{S}'$ is a "full subcategory" of $\underline{S}$ and $E': \underline{S}' \to \underline{B}$ is the restriction of E to $\underline{S}'$. A <u>full subcategory</u> $\underline{S}'$ of $\underline{S}$ is a restriction of $\underline{S}$ to a class of objects and all $\underline{S}$-morphisms between these objects (cf. 12.1).
Objects of the system category $\underline{S}$ will be called <u>systems</u> and systems S_1, S_2 are called <u>equivalent</u> if they have the same behavior, i.e. $E(S_1) = E(S_2)$.

<u>Remark</u>: In fact a systematic is the same as a functor having a partially ordered class as codomain such that all the three examples given in 3.1 are already examples of systematics. Moreover restricting the class of automata to observable or reduced automata we get examples of subsystematics which will be important for our further considerations.

<u>3.2 Reduced and Minimal Systems</u>: Now we are able to formulate the notions of "minimal" and "reduced" systems in our general setting of systematics. According to 2.4 all these notions including "observable" coincide in the deterministic case but in the nondeterministic case we have in fact differences (cf. 2.7). First of all it is easy to generalize the notion "reduced", given in 2.4, to systematics. We only have to replace surjective automata morphisms by a suitable class $\mathfrak{E}$ of "epimorphisms" and isomorphisms of automata by "isomorphisms" in our system category $\underline{S}$. An epimorphism $f: S \to S'$ is defined by the property that for all morphisms $f_1, f_2: S' \to S''$ with $f_1 \circ f = f_2 \circ f$ we already have $f_1 = f_2$ (cf. 12.3). Similar to the case $\underline{S} = \underline{Set}$ in which epimorphisms are exactly surjective functions we assume for the class $\mathfrak{E}$ that it is closed under composition and that $g \circ f \in \mathfrak{E}$ implies $g \in \mathfrak{E}$. Isomorphisms are morphisms $f: A \to B$ having an "inverse" morphism $f^{-1}: B \to A$ which satisfies $f^{-1} \circ f = id_A$ and $f \circ f^{-1} = id_B$ (cf. 12.3).
In order to find a suitable notion for "minimal" in a systematic, we refer back to the minimization problem in 2.2 which corresponds to minimization in the sense of minimal realiza-

tion of the behavior object E(A) . In fact a corresponding
suitable characterization of a minimal automaton is given in
part (iii) of the theorem in 2.4 which at once can be generalized to systematics: S' will be called minimal if for all
systems S with $E(S) \subseteq E(S')$ there is a unique morphism
of systems $f:S \to S'$ belonging to 𝕮 in the case $E(S) = E(S')$.
Since morphisms in 𝕮 correspond to surjective functions
minimal systems will have minimal cardinality of states in
some sense. Although observability cannot be formulated in
systematics we will see that observability of automata corresponds to minimality and weak minimality in the deterministic and nondeterministic case respectively.

<u>General Assumption</u>: In all our considerations $\underline{S} = (\underline{S},\underline{B},E)$
will be a systematic in the sense of the above definition ,
and 𝕮 will be a class of epimorphisms in $\underline{S}$ including all
isomorphisms, which is closed under composition, and for
$g \circ f \in 𝕮$ we assume $g \in 𝕮$. Moreover $\underline{S}'$ will denote a subsystematic $\underline{S}' = (\underline{S}',\underline{B},E')$ of $\underline{S}$.

<u>Definition</u> (R e d u c e d - M i n i m a l) : A system
S' is called <u>reduced</u> if each reduction $f:S' \to S$ to an
arbitrary system S is already an isomorphism. A <u>reduction</u>
$f:S' \to S$ is a morphism of systems which belongs to the
class 𝕮 and satisfies $E(S') = E(S)$. A system S' is
called <u>minimal</u> if for all systems S satisfying $E(S) \subseteq E(S')$
there is exactly one morphism $f:S \to S'$ in $\underline{S}$ which is
already a reduction in the case of equal behavior
$E(S) = E(S')$.

<u>Remark</u>: There is also a slightly weaker notion of minimality
assuming only the existence of a unique morphism $f:S \to S'$
in the case $E(S) = E(S')$ (cf. [3,5]) but this condition
does not imply minimal cardinality of states in general.
Confer [30] for counterexamples.

<u>Proposition</u>: Minimal systems are reduced and equivalent minimal systems are isomorphic.

<u>Proof</u>: Given a minimal system S' and a reduction $f:S' \to S$ we have $E(S')=E(S)$. Hence there is a unique morphism $f':S \to S'$ because S' is minimal. Now $f'\circ f$ and $id_{S'}$ are morphisms from S' to S'. Since S' is minimal we have $f'\circ f = id_{S'}$. Finally we get $f\circ f'\circ f = f\circ id_{S'} = id_S \circ f$ and hence $f\circ f' = id_S$ because f is a reduction and hence an epimorphism. In order to prove the second condition we assume to have two minimal systems S' and S satisfying $E(S')=E(S)$. Hence by minimality of S there is a reduction $f:S' \to S$ which is already an isomorphism because S' is also reduced.

∎

<u>3.3 Subsystematics</u>: Now we want to define suitable subsystematics of $\underline{S}$. Regarding the reduction problem in 2.2 we want to have for each automaton A not only a reduced automaton $\bar{A}$ but also a reduction $e:A \to \bar{A}$ which satisfies the universal property given in (iv) of 2.4. Hence in order to define a reduced subsystematic we will also require this universal property. The definitions of minimal and realizing subsystematics are straightforward.

<u>Definition</u> (R e d u c e d , M i n i m a l a n d R e a l i z i n g S u b s y s t e m a t i c s):
A subsystematic $\underline{S}'$ of $\underline{S}$ is called

(i)   <u>minimal</u>, if all systems S' in $\underline{S}'$ are minimal

(ii)   <u>realizing</u>, if for each behavior object B in $\underline{B}$ there is at least one system S' in $\underline{S}'$ such that $E'(S')=E(S')=B$

(iii) <u>reduced</u>, if all systems S' in $\underline{S}'$ are reduced and for each system S in $\underline{S}$ there is a reduction $u(S):S \to R(S)$ from S to a reduced system $R(S)$ in $\underline{S}'$ such that for all morphisms $f:S \to S'$ from S to a reduced system S' in $\underline{S}'$ there is a unique morphism f':R(S) → S' in (3.3.1) satisfying f'∘u(S) = f .

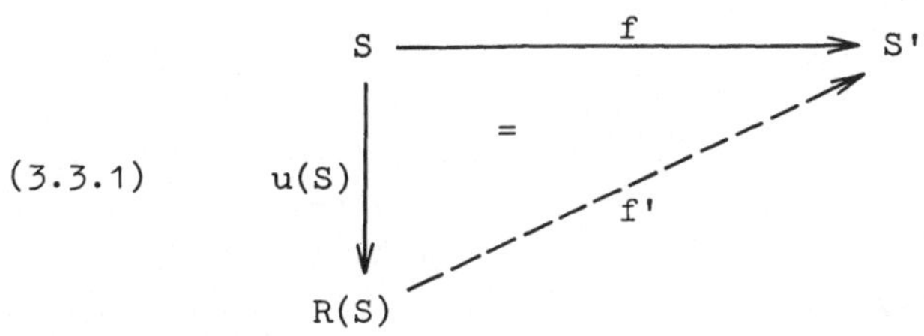

Remark: Given such "universal morphisms" u(S):S → R(S) the
category $\underline{S}$' is called reflexive subcategory of $\underline{S}$. This
"universal problem" is closely related to a "special problem"
which is stated in 3.4,3. A general categorical correspondence between universal and special problems is given in
[23,68].

3.4 Theorem (R e d u c t i o n) : Given a reduced sub-
systematic $\underline{S}$' of $\underline{S}$ we have the following properties:

1. There is a reduction functor $R:\underline{S} → \underline{S}$' satisfying
   E'∘R = E, i.e. R assigns to each system S∈$\underline{S}$ an equivalent reduced system R(S)∈$\underline{S}$'.

2. The reduced system R(S) of S is uniquely determined
   up to isomorphism by the property that there is a reduction u(S):S → R(S) from S to a reduced system R(S)
   in $\underline{S}$'.

3. The reduction process is decomposable in the following
   sense: For each reduction $f:S → S_1$ (S_1 not necessarily
   reduced) there is a unique reduction $f_1:S_1 → R(S)$ such
   that $f_1∘f = u(S)$.

4. Defining systems S_1 and S_2 to be R-equivalent if
   there is a (possibly) alternating chain of reductions
   between them, we have the following property: Systems are

R-equivalent iff the reduced systems are isomorphic. Moreover R-equivalence implies equivalence, but in general not vice versa.

5. $\underline{S}$ realizing implies $\underline{S}'$ realizing.

<u>Proof</u>: ı. Given a morphism $f:S \to S_1$ of systems the morphism $R(f):R(S) \to R(S_1)$ in (3.4.1) is uniquely defined to satisfy $R(f) \circ u(S) = u(S_1) \circ f$ using the universal property of $u(S)$ given in 3.3 (iii) and the fact that $u(S_1) \circ f:S \to R(S_1)$ is a morphism to a reduced system.

$$(3.4.1)$$

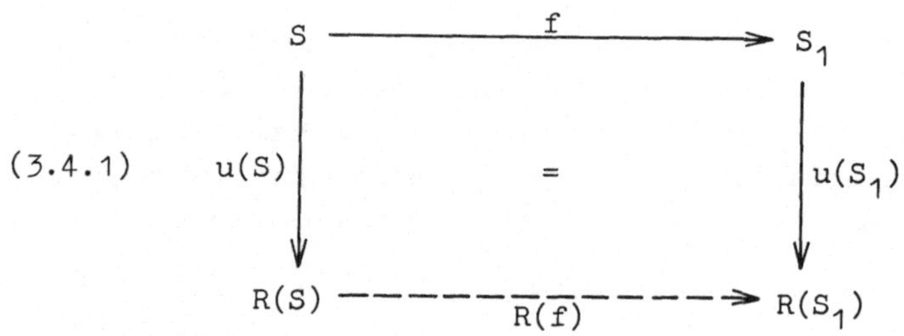

Given $g:S_1 \to S_2$ it is easy to show that $R(g) \circ R(f) = R(g \circ f)$ and $R(\mathrm{id}_S) = \mathrm{id}_{R(S)}$ using the uniqueness properties. Hence $R:\underline{S} \to \underline{S}'$ is a functor, called reduction functor. Since $u(S):S \to R(S)$ is a reduction we have $E(S) = E(R(S)) = E'(R(S)) = E' \circ R(S)$ and hence $E = E' \circ R$ because $\underline{B}$ is partially ordered.

2. Given a reduction $f:S \to S'$ from S to another reduced system S' in $\underline{S}'$ there is a unique morphism $f':R(S) \to S'$ satisfying $f' \circ u(S) = f$ by 3.3 (iii). Now $f \in \mathfrak{C}$ implies $f' \in \mathfrak{C}$ by the assumption in 3.2. Hence f' is a reduction and thus an isomorphism because $R(S)$ is reduced.

3. Given a reduction $f:S \to S_1$ we have $R(f) \circ u(S) = u(S_1) \circ f \in \mathfrak{C}$ by the first part of the proof. Hence we have $R(f) \in \mathfrak{C}$ and $E(R(S)) = E(S) = E(S_1) = E(R(S_1))$. Hence $R(f)$ is a reduc-

tion and thus an isomorphism because $R(S)$ is reduced.
Now we take $\qquad f_1 := (R(f))^{-1} \circ u(S_1) \qquad$ satisfying
$f_1 \circ f = (R(f))^{-1} \circ u(S_1) \circ f = (R(f))^{-1} \circ R(f) \circ u(S) = u(S)$.

4. Given R-equivalent systems S_1 and S_2 each reduction
in the chain from S_1 to S_2 induces an isomorphism between the corresponding reduced systems and hence we have an
isomorphism between $R(S_1)$ and $R(S_2)$. Vice versa we have
the chain

$$S_1 \xrightarrow{\;u(S_1)\;} R(S_1) \quad \widetilde{=} \quad R(S_2) \xleftarrow{\;u(S_2)\;} S_2$$

of reductions. Finally $R(S_1) \widetilde{=} R(S_2)$ implies $E(S_1) =$
$= E(R(S_1)) \widetilde{=} E(R(S_2)) = E(S_2)$, since E is a functor, and
hence equality because $\underline{B}$ is partially ordered. Thus S_1
and S_2 are equivalent. Vice versa there are equivalent reduced nondeterministic automata which are not isomorphic
(cf. 2.7).

5. Given a behavior B which is realized by S in $\underline{S}$ it
is also realized by $R(S)$ in $\underline{S}'$ because $E(R(S)) = E(S) = B$
by 1.
∎

<u>3.5 Adjoint Functors</u>: In 3.4,1 it is shown that the univer-
sal properties of the morphisms $u(S):S \to R(S)$ defined in
3.3 (iii) lead to a construction of a functor $R:\underline{S} \to \underline{S}'$.
In fact this is a general categorical construction leading
to the notion of adjoint functors (cf. 12.7). In our case
$f':R(S) \to S'$ can be replaced by $J(f'):J(R(S)) \to J(S')$
where J only denotes the inclusion from the category $\underline{S}'$
to $\underline{S}$ which is a trivial functor $J:\underline{S}' \to \underline{S}$, called inclusion functor. Thus we have a pair of functors $R:\underline{S} \to \underline{S}'$
and $J:\underline{S}' \to \underline{S}$ in opposite directions which are called
adjoint functors. More precisely R, corresponding to
$S:\underline{G} \to \underline{K}$ in 12.7, is called left adjoint to J which corresponds to $T:\underline{K} \to \underline{G}$. For most of the following constructions
we will need the dual process.

<u>Definition</u>: Given a functor $S:\underline{G} \to \underline{K}$ the couniversal prob-
lem of S is said to be solvable if for each object K in $\underline{K}$

there is an object T(K) in $\underline{G}$ and a couniversal morphism
v(K):ST(K) → K such that for all G in $\underline{G}$ and all morphisms h:S(G) → K in $\underline{K}$ there is exactly one morphism
f:G → T(K) in $\underline{G}$ satisfying v(K)∘S(f) = h :

(3.5.1)

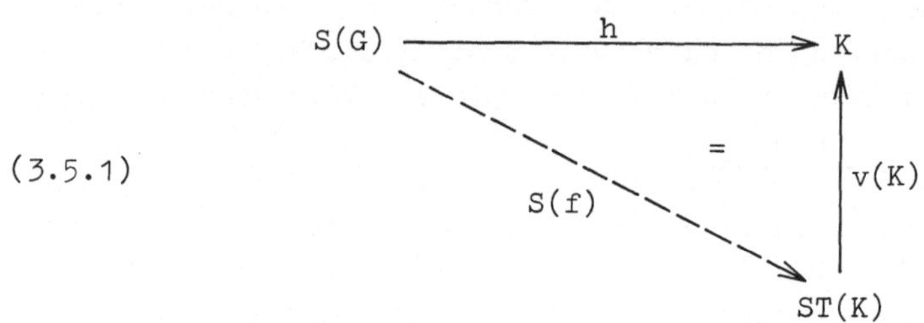

Dually to 3.4,1 T can be extended to a functor T:$\underline{K}$ → $\underline{G}$
which is now called right adjoint functor to S:$\underline{G}$ → $\underline{K}$.
Moreover it is easy to show that T is right adjoint to S
iff S is left adjoint to T , written S ⊣ T.
In fact the inclusion J:$\underline{S}$' → $\underline{S}$ in the above example is
right adjoint to the reduction functor R:$\underline{S}$ → $\underline{S}$' and now we
will be concerned with a right adjoint M*:$\underline{B}$ → $\underline{S}$ to the
behavior functor E:$\underline{S}$ → $\underline{B}$ which in addition is right inverse to E , i.e. E∘M* = Id$_{\underline{B}}$. Due to J. A. G o g u e n
such a functor will be called minimal realization and E is
called to satisfy the Minimal Realization Principle in this
case (cf. [44,45]). We use the symbol M* to avoid confu-
sion with the construction M(A) of the machine function
in 2.1.

<u>3.6 Minimal Realization</u>: First we state the Minimal Realiza-
tion Principle in a slightly different form.

<u>Definition</u>: A systematic $\underline{S}$ = ($\underline{S}$,$\underline{B}$,E) is called to satisfy
the <u>Minimal Realization Principle</u> if there is a right adjoint
and right inverse functor M*:$\underline{B}$ → $\underline{S}$ to the behavior functor
E:$\underline{S}$ → $\underline{B}$ such that the universal morphisms are reductions.

In other words there is a functor $M^*:\underline{B} \to \underline{S}$ such that
$E \circ M^* = Id_{\underline{B}}$ and for all systems S and behavior objects B
satisfying $E(S) \subseteq B$ there is a unique morphism $f:S \to M^*(B)$
in (3.6.1) which is a reduction in the case of equality.

(3.6.1)

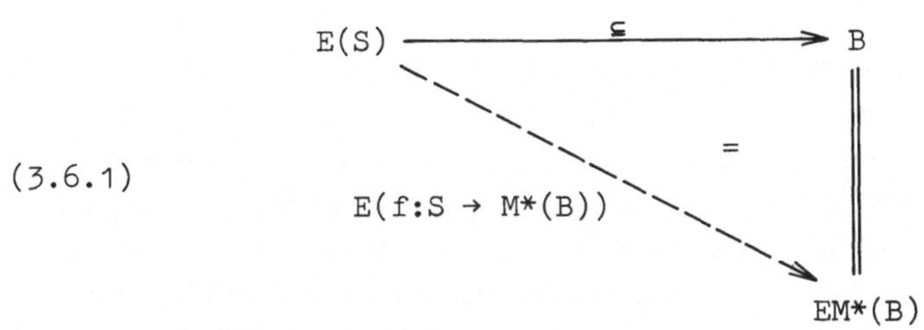

$M^*:\underline{B} \to \underline{S}$ is called <u>minimal realization functor</u> and $M^*(B)$
<u>minimal realization</u> of $B \in \underline{B}$ where $M^*(B)$ is minimal in the
sense of the definition in 3.2 by (3.6.1).

<u>Theorem</u> (M i n i m a l   R e a l i z a t i o n) :

1. Given a systematic $\underline{\underline{S}} = (\underline{S},\underline{B},E)$ there exists a minimal and
   realizing subsystematic $\underline{\underline{S}}'$ of $\underline{\underline{S}}$ iff $\underline{\underline{S}}$ satisfies the
   Minimal Realization Principle.

Given a minimal and realizing subsystematic $\underline{\underline{S}}' = (\underline{S}',\underline{B},E')$
of $\underline{\underline{S}}$ we have:

2. The class of objects in $\underline{S}'$ is exactly the class of all
   minimal systems and coincides with the class of all systems which are isomorphic to minimal realizations $M^*(B)$
   of behavior objects B , provided that $\underline{S}'$ is closed
   under isomorphisms, i.e. each object in $\underline{S}$, isomorphic to
   an object in $\underline{S}'$, belongs already to $\underline{S}'$. In this case $\underline{S}'$
   is uniquely determined.

3. $\underline{\underline{S}}'$ is already reduced and ,regarding M^* as functor from
   $\underline{B}$ to $\underline{S}'$, $M^* \circ E:\underline{S} \to \underline{S}'$ is already a reduction functor R.
   Thus reduction coincides with minimization.

4. Systems are equivalent iff they are R-equivalent or iff
the reduced systems are isomorphic.

5. Systems are minimal iff they are reduced. Moreover reduction, minimization and minimal realization are unique up
to isomorphism.

Proof: 1. Given a minimal and realizing subsystematic $\underline{S}'$
and a behavior object B there is a system S' in $\underline{S}'$,
which is minimal of course, such that $E(S') = B$. Defining
$M*(B) = S'$, $M*(B)$ is minimal and we have $EM*(B) = E(S') = B$.
Moreover, $E(S) \subseteq B = EM*(B)$ for an arbitrary system S in $\underline{S}$
implies that there is a unique morphism $f:S \to M*(B)$ which
is a reduction in the case $E(S) = B$ because $M*(B)$ is
minimal. Given $B \subseteq B'$ and hence $EM*(B) \subseteq EM*(B')$ there
is a unique morphism $f:M*(B) \to M*(B')$ such that $M*$ be-
comes a functor satisfying $E \cdot M* = Id_{\underline{B}}$. Hence $\underline{S}$ satisfies
the Minimal Realization Principle.

Vice versa $\underline{S}'$ is defined to consist of all systems which
are isomorphic to minimal realizations $M*(B)$ for arbitrary
B in $\underline{B}$. Using the properties of $M*$ it is easy to see
that $M*(B)$ is minimal. Hence $\underline{S}'$ is minimal and realizing.

2. By assumption, $\underline{S}'$ is a subclass of all minimal systems.
Vice versa, given a minimal system S, $M*E(S) \in \underline{S}'$ is equivalent to S and minimal. Hence S is isomorphic to $M*E(S)$
by the proposition in 3.2 and belongs to $\underline{S}'$ by assumption
on $\underline{S}'$. The second part of the assertion is obvious, using
the fact that the property to be minimal is closed under
isomorphisms.

3. Defining $R(S) := M*E(S)$ we have $ER(S) = EM*E(S) = E(S)$
and $R(S)$ is already minimal and hence reduced (cf. 3.2).
Moreover there is a unique reduction $u(S):S \to R(S)$ since
$E(S) = EM*E(S) = ER(S)$ and $R(S)$ is minimal. Finally given a
morphism $f:S \to S'$ with S' in $\underline{S}'$ we have $ER(S) = E(S) \subseteq$
$\subseteq E(S')$. Thus using the minimality of S' in $\underline{S}'$ there is
a unique morphism $f':R(S) \to S'$ such that $f' \cdot u(S) = f$.

4. By 3.4,4 it remains to show that the reduced systems of
equivalent systems S and S_1 are isomorphic. Now R(S)
and $R(S_1)$ are minimal by 3. and equivalent and hence
isomorphic by 3.2.

5. Given a reduced system S the reduction $u(S):S \to R(S)$
which exists by 3. is already an isomorphism. Hence
$S \cong R(S) = M*E(S)$ is minimal. The converse has already been
shown in 3.2. Moreover uniqueness of reduction, minimization
and minimal realization up to isomorphism is a direct consequence of the fact that equivalent minimal systems are isomorphic (cf. 3.2).   ∎

Applications of this theorem will be given in the following
chapters 5, 7 and 9.

<u>3.7 Finite Cardinality</u>: In 3.2 we have remarked that minimal
systems in some sense have minimal cardinality of states.
Now we want to be more precise: We assume that there is a
cardinality function card assigning to each system a cardinal
number, for example the cardinality of states or the dimen-
sion of vectorspaces. Moreover given a reduction $f:S \to S_1$
we assume $card(S_1) \leq card(S)$ and if card(S) is equal to
$card(S_1)$ and finite we assume that f is already an isomorphism. This is a typical property of functions between
finite sets or finite dimensional vector spaces.

Now each minimal system S' has minimal cardinality with
respect to all equivalent systems S , because E(S) = E(S')
implies that there is a reduction $f:S \to S'$ and hence
$card(S') \leq card(S)$. Vice versa each system S', having finite
cardinality card(S') satisfying $card(S') \leq card(S)$ for
all equivalent systems S , is in fact minimal provided that
there is at least one minimal system S_1 equivalent to S' :
Since $card(S') \leq card(S_1)$ by assumption and $card(S_1) \leq$
$\leq card(S')$ by minimality of S_1 we have equality. Hence
there is a reduction $f:S' \to S_1$ which is already an isomorphism using the second property of the cardinality func-

tion. Hence S' is minimal, too.

We state a weaker version of our result:

<u>Theorem</u>: Given a minimal and realizing or minimal and re-
duced subsystematic $\underline{S}'$ of $\underline{S}$ and a cardinality function
in the above defined sense a system S' with finite cardi-
nality is minimal iff we have card(S') $\leq$ card(S) for all
systems S equivalent to S'.

4. Behavior of Automata in Closed Categories: The Deterministic Case

The aim of this chapter is to generalize the construction of
the extended output function, the machine function and the
input-output behavior of deterministic automata to those in
monoidal categories which are introduced in chapter 1.
As motivated in the introduction to chapter 2 we use the
universal constructions given in 2.1 to formulate the theory
in categorical terms. Thus we need "coproducts" being a generalization of disjoint unions in the set case, and for the
construction of the machine function we assume that the category $(\underline{K},\otimes)$ is "closed" which means that the tensor product $-\otimes K:\underline{K} \to \underline{K}$ has a right adjoint functor for all objects
K in $\underline{K}$. The presence of "closed categories" is specific
for automata of deterministic type such as deterministic,
partial, bilinear and topological automata which are studied
as main examples, confer 4.9 for others. Automata in pseudoclosed categories corresponding to the nondeterministic case
will be introduced in chapter 6. Finally in order to construct the behavior we have to assume that there is an
"image factorization" in our closed category $(\underline{K},\otimes)$.
All these categorical notions will be motivated and introduced together with the automata theoretic concepts. Main
results of this chapter are characterizations of automata
using their machine functions (cf.4.5) and of the behavior
(cf. 4.8). Reduction, minimization and realization will be
studied in chapter 5, using the general concepts which are
introduced in chapter 3.

<u>4.1 General Assumptions</u>: In all our considerations $(\underline{K},\otimes)$
will denote a monoidal category, $\underline{K}\text{-}\underline{Aut}$ the category of
automata in $(\underline{K},\otimes)$ and $\underline{K}\text{-}\underline{Medv}$ the corresponding category of

Medvedev-automata, cf. 1.9, 1.11 and 1.14 respectively.
Additional assumptions for $(\underline{K},\otimes)$ will be introduced later on:

(i) $(\underline{K},\otimes)$ has countable coproducts (cf. 4.2)

(ii) $(\underline{K},\otimes)$ is closed with right adjoint $<K,->:\underline{K} \to \underline{K}$
      to $-\otimes K:\underline{K} \to \underline{K}$ for each object K in $\underline{K}$ (cf. 4.3)

(iii) $(\underline{K},\otimes)$ has an $\mathfrak{C}$-$\mathfrak{M}$-factorization and canonical
      representatives in $\mathfrak{M}$ (cf. 4.6).

<u>4.2 Extended Output Morphism and Coproducts</u>: In order to
generalize the construction of the extended output function
as given in 2.1 for deterministic automata we need an analo-
gue to the disjoint union $\underset{n\in\mathbf{N}}{\uplus}$ of sets which is given by the
categorical notion of countable coproducts.

<u>Definition</u>: Given a family K_n ($n\in\mathbf{N}$) of objects in $\underline{K}$ an
object $\bar{K}$, written $\underset{n\in\mathbf{N}}{\amalg} K_n$, together with "injections"
$u_n:K_n \to \bar{K}$ is called <u>coproduct</u> of the family $[K_n]_{n\in\mathbf{N}}$ if they
have the following universal property: For all objects L
in $\underline{K}$ and all families $f_n:K_n \to L$ ($n\in\mathbf{N}$) of morphisms there
is a unique morphism $f:\bar{K} \to L$ in (4.2.0) such that $f\circ u_n = f_n$
for all $n\in\mathbf{N}$ (cf. 12.4).

(4.2.0)

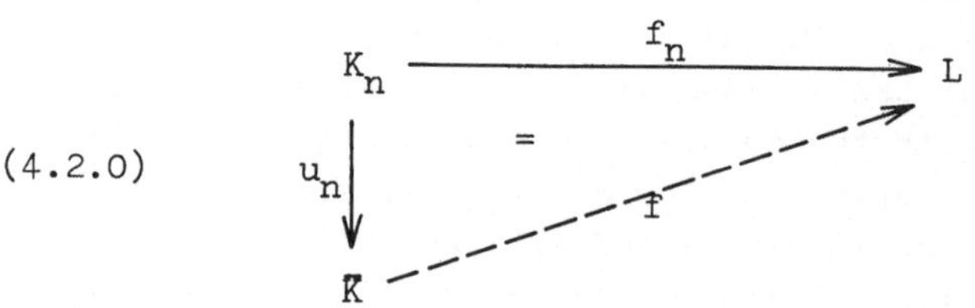

The morphism f is called to be induced by $[f_n]_{n\in\mathbf{N}}$,
written $f = \langle f_n \rangle_{n\in\mathbf{N}}$.
<u>Remark</u>: The coproduct object $\bar{K}$ is uniquely determined up
to isomorphism (cf. 12.9).

<u>Examples</u>: In the category <u>Set</u> the disjoint union $\bigcup\limits_{n\in\mathbb{N}} K_n$ of a family K_n $(n\in\mathbb{N})$ together with the inclusions $u_n:K_n \to \bigcup\limits_{n\in\mathbb{N}} K_n$ have these universal properties of a coproduct. In fact given a family of functions $f_n:K_n \to L$ we define $f: \bigcup\limits_{n\in\mathbb{N}} K_n \to L$ by $f(x) = f_n(x)$ for $x\in K_n$. Thus we have $f\circ u_n(x) = f(x) = f_n(x)$ and hence $f\circ u_n = f_n$. Vice versa this condition implies the above definition of f and hence the uniqueness of f .

Exactly the same construction can be used in the categories <u>PD</u> of partial functions and <u>Top</u> of topological spaces. In <u>PD</u> f is only partially defined and in <u>Top</u> the coproduct is the topological sum $\sum K_n$ of the K_n such that f is continuous iff all of the functions f_n are continuous. Finally in <u>Mod</u>$_R$ the coproduct is given by the direct sum $\oplus K_n$.

<u>Assumptions</u>: Henceforth we will assume that $(\underline{K},\otimes)$ has <u>countable coproducts</u> meaning that each countable family K_n $(n\in\mathbb{N})$ of objects in $\underline{K}$ has a coproduct (cf. 4.1 (i)). Moreover we will assume that for each coproduct $\bar{K}$ with injections $u_n:K_n \to \bar{K}$ and each object K in $\underline{K}$ $K\otimes\bar{K}$ again is a coproduct of the family $K\otimes K_n$ with injections $K\otimes u_n:K\otimes K_n \to K\otimes\bar{K}$. This means that the functor $K\otimes-:\underline{K} \to \underline{K}$ preserves coproducts. In fact this will turn out to be a consequence of assumption 4.1 (ii).

Now we are able to construct I^+ and $1^+:S\otimes I^+ \to 0$ in exactly the same way as given in 2.1.

<u>Constructions and Definitions</u>: Given an input object I in $\underline{K}$ define $I^1 = I$ and $I^{n+1} = I\otimes I^n$ for all $n\in\mathbb{N}$. Now the coproduct of the family I^n $(n\in\mathbb{N})$, written $I^+ = \coprod\limits_{n\in\mathbb{N}} I^n$ with injections $i_n:I^n \to I^+$ $(n\in\mathbb{N})$, is called <u>free semigroup</u> or <u>strings</u> on I .

Moreover given an automaton $A = (S,d,1)$ in $\underline{K}$-<u>Aut</u> we construct a family $1_n:S\otimes I^n \to 0$ by

(4.2.1) $\quad 1_1 = 1$ and $1_{n+1} = 1_n \circ (d\otimes I^n)$ for all $n\in\mathbb{N}$.

This leads to a unique morphism $1^+ : S \otimes I^+ \to 0$, called __extended (last) output__, such that

$$(4.2.2) \qquad 1_n = 1^+ \circ (S \otimes i_n) \quad \text{for all } n \in \mathbb{N}$$

using the coproduct properties of $S \otimes I^+$.

In our examples I^+ is exactly the set of all non-empty strings on I and 1^+ assigns to each state and input string the corresponding last output symbol. Only in the case $(\underline{\mathrm{Mod}}_R, \otimes)$, I^+ is not the set of all strings on I but $1^+ : S \otimes I^+ \to 0$ can be interpreted in a similar way using the corresponding multilinear functions $\tilde{1}_n : S \times I \times I \times \cdots \times I \to 0$ of $1_n : S \otimes I^n \to 0$.

$$\underbrace{\qquad\qquad}_{n}$$

The construction of 1^+ leads to the following nice compatibility property of 1^+ which will be needed later on:

__Lemma:__ Let $i : I \otimes I^+ \to I^+$ be the inclusion defined by the family $i_{n+1} : I \otimes I^n \to I^+$, i.e.

$$(4.2.3) \qquad i_{n+1} = i \circ (I \otimes i_n) \quad \text{for all } n \in \mathbb{N},$$

then we have

$$(4.2.4) \qquad 1^+ \circ (S \otimes i) = 1^+ \circ (d \otimes I^+) : S \otimes I \otimes I^+ \to 0 .$$

__Proof:__ Using the uniqueness of a morphism out of the coproduct $S \otimes I \otimes I^+$ the assertion is equivalent to

$$1^+ \circ (S \otimes i) \circ (S \otimes I \otimes i_n) = 1^+ \circ (d \otimes I^+) \circ (S \otimes I \otimes i_n) \quad \text{for all } n \in \mathbb{N}.$$

In fact we have:

$$\begin{aligned}
1^+ \circ (d \otimes I^+) \circ (S \otimes I \otimes i_n) &= 1^+ \circ (S \otimes i_n) \circ (d \otimes I^n) \\
&= 1_n \circ (d \otimes I^n) \\
&= 1_{n+1} \\
&= 1^+ \circ (S \otimes i_{n+1}) \\
&= 1^+ \circ (S \otimes i) \circ (S \otimes I \otimes i_n) .
\end{aligned}$$

In the first step we have used the bifunctor properties of $\otimes$ and in the other steps the definitions of 1^+, 1_{n+1}, 1^+ and i respectively. $\blacksquare$

<u>4.3 Machine Morphisms and Closed Categories</u>: Our next step
is the generalization of the machine function $M(A):S \to <I^+,O>$
which assigns to each state $s \in S$ the corresponding input-
output function $M(s) = 1^+(s,-):I^+ \to O$. In fact we will use
the universal construction given in the remark of 2.1.
This is a bijective correspondence between morphisms
$f:S \times I^+ \to O$ and $f':S \to <I^+,O>$ using the object $<I^+,O>$ of
all morphisms from I^+ to O . More precisely this property
is characterized by the fact that $-\times I^+:\underline{Set} \to \underline{Set}$ has a
right adjoint functor $<I^+,->:\underline{Set} \to \underline{Set}$ (cf. 3.5). In fact
we have corresponding properties in the categories $(\underline{PD},\times)$,
$(\underline{Mod}_R,\otimes)$ and $(\underline{Top},\otimes)$ which are needed in our examples.
Such categories are called closed:

<u>Definition</u>: A monoidal category $(\underline{K},\otimes)$ is called <u>closed</u> if
for all $K \in \underline{K}$ the functor $-\otimes K:\underline{K} \to \underline{K}$ has a right adjoint
$<K,->:\underline{K} \to \underline{K}$, i.e. for each object $O \in \underline{K}$ there is a morphism
$ev:<K,O>\otimes K \to O$, called <u>evaluation</u>, such that for all ob-
jects $S \in \underline{K}$ and all morphisms $f:S \otimes K \to O$ there is a unique
morphism $f':S \to <K,O>$ in (4.3.1) satisfying $ev \circ (f' \otimes K) = f$.

(4.3.1)

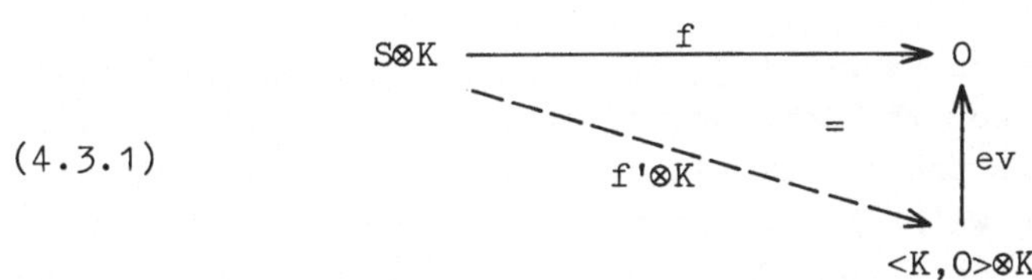

Especially the uniqueness property of f' implies for
$f',g':S \to <K,O>$

(4.3.2) $f' = g'$ iff $ev \circ (f' \otimes K) = ev \circ (g' \otimes K)$.

Moreover we will assume that a closed category is <u>symmetric</u>
in the sense that there is a canonical isomorphism $s_{AB}:A \otimes B \overset{\sim}{\to}$
$\overset{\sim}{\to} B \otimes A$ for each pair of objects A,B in $\underline{K}$ satisfying

64

$s_{BA} \circ s_{AB} = id_{A \otimes B}$ and which is "compatible" with right unit, left unit and associativity (cf. 1.9). The exact conditions are given in [59] chapter 7.

<u>Examples</u>: 1. First of all (<u>Set</u>,×) is a closed category. In fact this has already been shown in the remark of 2.1, if we replace I^+ by an arbitrary set K . Moreover the symmetry s_{AB}:A×B $\overset{\sim}{\to}$ B×A of the cartesian product is given by $s_{AB}(a,b) = (b,a)$ of course.

2. In the case (<u>PD</u>,×) we take <K,O> to be the set of all non-empty partial functions from K to O and for ev:<K,O>×K → O we take the partial evaluation function which for $(g,a) \in$ <K,O>×K is defined by ev(g,a) = g(a) iff g(a) is defined and not defined otherwise. From these definitions it is only an exercise to show the universal properties for (<u>PD</u>,×) to be closed.

3. (<u>Mod</u>$_R$,⊗) is also closed taking for <K,O> the set of all R-linear functions which is in fact an R-module. Now ev:<K,O>⊗K → O is induced by the bilinear evaluation function ev:<K,O>×K → O due to the universal property given in 1.6.

4. Finally the category (<u>Top</u>,⊗) is closed using the biproduct A⊗B and not the topological product (cf. 1.7). In this case <K,O> is the set of all continuous functions from K to O endowed with the topology of pointwise convergence, i.e. for each a∈K and O' open in O the set B(a,O') of all continuous functions f:K → O satisfying f(a)∈O' is open in <K,O> and the system B(a,O') for all a∈K, O' open in O is a subbasis of the topology of <K,O>. Now using the definition of the bitopology given in 1.7 it is easy to show that the evaluation function ev:<K,O>⊗K → O is continuous, and given a continuous function f:S⊗K → O the functions f':S → <K,O> defined by f'(s)(a) = f(s,a) and f'⊗K are again continuous.

General Assumptions: For this and the next chapter we will
assume that $(\underline{K},\otimes)$ is closed (cf. 4.1 (ii)). Thus for all
K $-\otimes K$ is a left adjoint functor and preserves coproducts
(cf. 12.9). Moreover by symmetry of $\otimes$ we also have this
property for $K\otimes-$ and thus the second part of the assumption in 4.2.

Now we are able to construct the machine morphism in the
same way as given in 2.1.

Construction and Definition: Given an automaton $A = (S,d,l)$
in $(\underline{K},\otimes)$ with extended output $l^+:S\otimes I^+ \to 0$ there is a
unique morphism $M(A):S \to {<}I^+,0{>}$ in (4.3.3), called
machine morphism of A , such that $ev\circ(M(A)\otimes I^+) = l^+$.

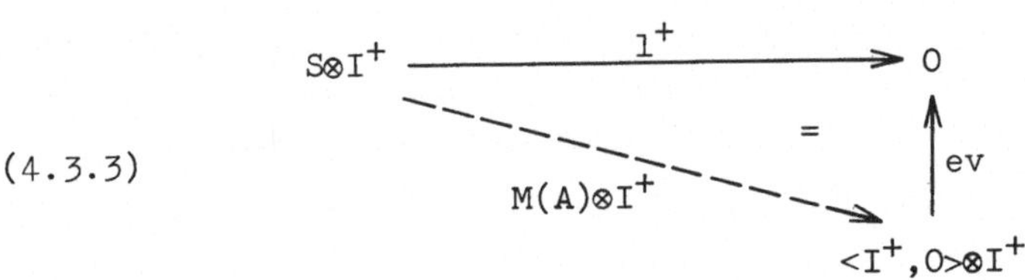

In all our examples $M(A)$ assigns to each state the corresponding input-output morphism $l^+(s,-):I^+ \to 0$ of this
state. The functions $M(A)$ and $l^+(s,-)$ are deterministic,
partial, R-linear and continuous in the examples $(\underline{Set},\times)$,
$(\underline{PD},\times)$, $(\underline{Mod}_R,\otimes)$ and $(\underline{Top},\otimes)$ respectively.

Remark: Our above definition of the machine morphism is
based on the fact that $-\otimes I^+:\underline{K} \to \underline{K}$ has a right adjoint
functor $<I^+,->:\underline{K} \to \underline{K}$ and we get the machine morphism as a
morphism $M(A):S \to {<}I^+,0{>}$ in $\underline{K}$. In fact, as shown in
lemma 2.3 for the deterministic case, $M(A)$ is already an
Medvedev-automata morphism from (S,d) to $({<}I^+,0{>},L)$.
In other words starting with a $\underline{K}$-morphism $l:S\otimes I \to 0$ we get

a $\underline{K}$-$\underline{Medv}$-morphism $g:(S,d) \to (<I^+,O>,L)$ and vice versa.
Using the forgetful functor $V:\underline{K}$-$\underline{Medv} \to \underline{K}$ defined by
$V(S,d) = S$ and $V(f) = f$ we in fact have shown in lemma 2.3
for $(\underline{K},\otimes) = (\underline{Set},\times)$ that the composite functor V-$\otimes I :=$
$= (-\otimes I) \circ V:\underline{K}$-$\underline{Medv} \to \underline{K}$ has a right adjoint functor
$<I^+,->:\underline{K} \to \underline{K}$-$\underline{Medv}$ assigning to the object O the Medvedev-
automaton $(<I^+,O>,L)$. As motivated in chapter 2 this ad-
junction, which now will be established for arbitrary closed
categories $(\underline{K},\otimes)$, is most important for reduction, minimization and behavior characterization of automata.

<u>4.4 Lemma</u>: Let $V:\underline{K}$-$\underline{Medv} \to \underline{K}$ be the forgetful functor from
the category of Medvedev-automata in $(\underline{K},\otimes)$ to the closed
category $(\underline{K},\otimes)$. Then the composite functor
V-$\otimes I:\underline{K}$-$\underline{Medv} \to \underline{K}$ has a right adjoint $<I^+,->:\underline{K} \to \underline{K}$-$\underline{Medv}$.

More detailed we have for each object O in $\underline{K}$ a $\underline{K}$-Medvedev-
automaton $(<I^+,O>,L)$ where the left-shift morphism
$L:<I^+,O>\otimes I \to <I^+,O>$ is the unique adjoint morphism of
$$<I^+,O>\otimes I \otimes I^+ \xrightarrow{\ <I^+,O>\otimes i\ } <I^+,O>\otimes I^+ \xrightarrow{\ ev\ } O \ , \quad \text{i.e.}$$

$$(4.4.1) \qquad\qquad ev\circ(L\otimes I^+) = ev\circ(<I^+,O>\otimes i) \ ,$$

and $i:I\otimes I^+ \to I^+$ is defined in the lemma of 4.2.

Let
$$(4.4.2) \qquad ev_1 := <I^+,O>\otimes I \xrightarrow{\ <I^+,O>\otimes i_1\ } <T^+,O>\otimes I^+ \xrightarrow{\ ev\ } O$$

be the restriction of the evaluation, then there is for each
$(S,d)\in\underline{K}$-$\underline{Medv}$ and each $\underline{K}$-morphism $1:S\otimes I \to O$ a unique
$\underline{K}$-$\underline{Medv}$-morphism $g:(S,d) \to (<I^+,O>,L)$ in $(4.4.3)$ satisfying
$ev_1\circ(g\otimes I) = 1$.

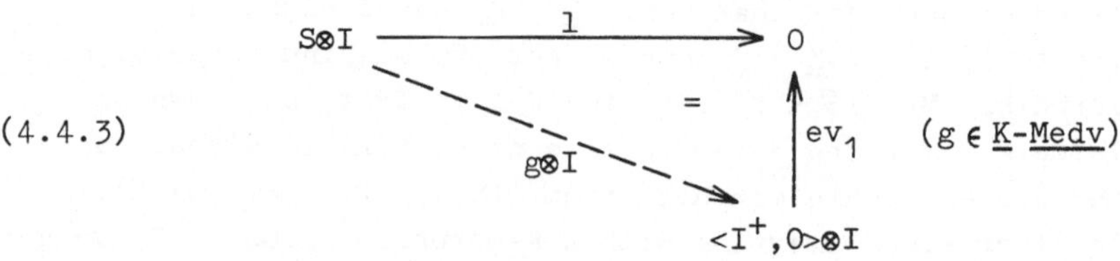

$$(4.4.3)$$

Moreover regarding $A = (S,d,l)$ as an automaton the above
constructed $\underline{K}$-morphism $g:S \to <I^+,O>$ is exactly the machine
morphism $M(A)$ of A .

<u>Proof</u>: Given $A = (S,d,l)$ and a $\underline{K}$-$\underline{Medv}$-morphism
$g:(S,d) \to (<I^+,O>,L)$ satisfying $ev_1 \circ (g \otimes I) = l$ we first
show $ev \circ (g \otimes I^+) = l^+$ and hence $g = M(A)$ by definition of
$M(A)$ in 4.3.3 which shows uniqueness of g .
Using the uniqueness properties of the coproduct $S \otimes I^+$ the
assertion $ev \circ (g \otimes I^+) = l^+$ is equivalent to

(4.4.4) $ev \circ (g \otimes I^+) \circ (S \otimes i_n) = l^+ \circ (S \otimes i_n) = l_n$ for all $n \in \mathbb{N}$

which will be shown by induction on n :
$n = 1$: $ev \circ (g \otimes I^+) \circ (S \otimes i_1) = ev \circ (<I^+,O> \otimes i_1) \circ (g \otimes I)$
$$= ev_1 \circ (g \otimes I) = l = l_1 \ .$$

In the first step we have used the bifunctor properties of $\otimes$
which will be abbreviated by $\otimes$ in the diagrams below.

The assertion (4.4.4) for $n+1$ $(n \geq 1)$ is given by the following diagram (4.4.5) which commutes by the references given
in the subdiagrams.

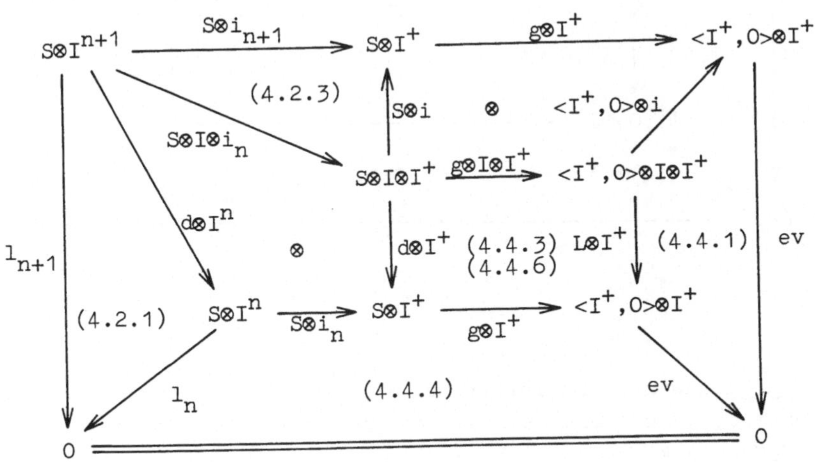

$$(4.4.5)$$

Now we define $g := M(A)$ given by (4.3.3) and have to show
(4.4.3) and that $M(A)$ is a Medvedev-automata morphism, i.e.

(4.4.6) $$M(A) \circ d = L \circ (M(A) \otimes I) .$$

First we have by (4.4.2), $\otimes$, (4.3.3) and (4.2.2) :

$$ev_1 \circ (M(A) \otimes I) = ev \circ (\langle I^+, O \rangle \otimes i_1) \circ (M(A) \otimes I)$$
$$= ev \circ (M(A) \otimes I^+) \circ (S \otimes i_1)$$
$$= 1^+ \circ (S \otimes i_1)$$
$$= 1_1 = 1 .$$

In order to get (4.4.6) it suffices to show that diagram (1)
in (4.4.7) is equalized by ev (cf.(4.3.2)). But this follows from the fact that the outer diagram in (4.4.7) is
commutative by (4.2.4) and the same is true for the remaining subdiagrams.

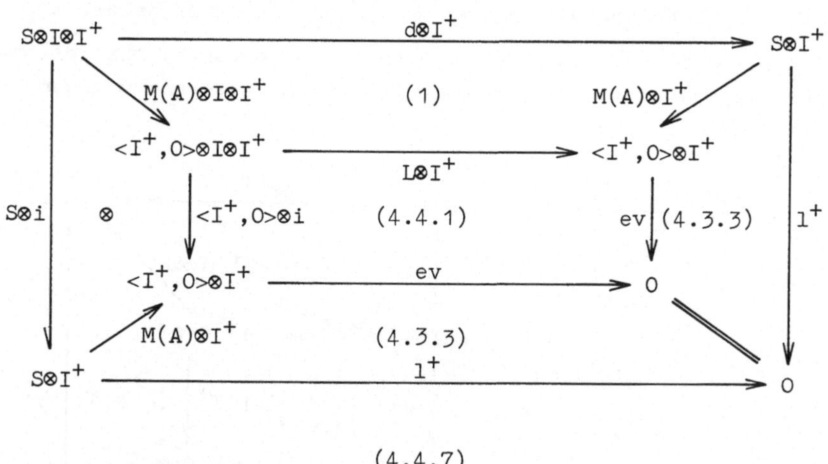

$$(4.4.7)$$

Remark: Another possibility of showing the adjunction of the
functors $V-\otimes I$ and $\langle I^+, -\rangle$ is to verify that the functor
$V : \underline{K}\text{-}\underline{Medv} \to \underline{K}$ has a right adjoint $\langle I*, -\rangle$, where $I* = I^+ \amalg U$
is the coproduct of I^+ with the unit object U of the
monoidal category $(\underline{K}, \otimes)$. Since we have already the adjunc-

tion of $-\otimes I$ and $<I,->$, by $(\underline{K},\otimes)$ being closed (cf. 4.3),
we get the above adjunction as a composition of adjoint
functors (cf. 12.7) using the natural isomorphisms
$<I^*,<I,->> \cong <I^*\otimes I,-> \cong <I^+,->$. This approach is given in
[28]. On the other hand, the adjunction of V and $<I^*,->$
is just the one we need in the case of Moore-automata
(cf. 1.14) because our output morphism $1:S\otimes I \to 0$ is replaced by $m:S \to 0$, which can be regarded as a $\underline{K}$ -morphism
$m:V(S,d) \to 0$. Hence the adjunction of V and $<I^*,->$ leads
to a machine morphism $M(A):S \to <I^*,0>$ which again is a
$\underline{K}$-$\underline{Medv}$-morphism.

In the Arbib-Manes approach of adjoint machines in [5] the
property of $V:\underline{K}$-$\underline{Medv} \to \underline{K}$ to have a right adjoint functor
is taken as an axiom. More precisely, in [5] they do not
start with a monoidal category $(\underline{K},\otimes)$ but with an arbitrary
category $\underline{K}$ where the functor $-\otimes I:\underline{K} \to \underline{K}$ is replaced by an
input functor $X:\underline{K} \to \underline{K}$. Hence the category $\underline{K}$-$\underline{Medv}$ is replaced by a category $\underline{Dyn}(X)$ with objects (S,d) such that
d is a $\underline{K}$ -morphism $d:X(S) \to S$ instead of $d:S\otimes I \to S$.
In this approach, starting on a higher level of generaliza-
tion than we do, it is possible in addition, to include
linear and tree automata.

Now lemma 4.4 leads to an important result characterizing
automata by their machine morphisms.

<u>4.5 Theorem</u> (C h a r a c t e r i z a t i o n   o f
K - A u t) : The category $\underline{K}$-$\underline{Aut}$ of automata in $(\underline{K},\otimes)$
is isomorphic to the following category $(\underline{K}$-$\underline{Medv} \downarrow <I^+,0>)$:
Objects of this category are $\underline{K}$-$\underline{Medv}$-morphisms $g:S \to <I^+,0>$
where (S,d) is an arbitrary and $(<I^+,0>,L)$ the fixed
Medvedev-automaton defined in lemma 4.4. Morphisms in
$(\underline{K}$-$\underline{Medv} \downarrow <I^+,0>)$ between the objects $g:S \to <I^+,0>$ and
$g':S' \to <I^+,0>$ in (4.5.1) are $\underline{K}$-$\underline{Medv}$-morphisms $f:S \to S'$
satisfying $g'\circ f = g$.

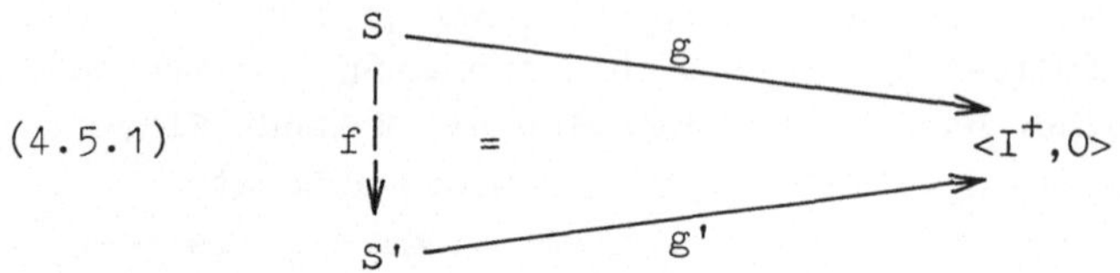

The isomorphism of categories

$$M : \underline{K\text{-}Aut} \xrightarrow{\sim} (\underline{K\text{-}Medv} \downarrow <I^+,O>)$$

assigns to each automaton $A \in \underline{K\text{-}Aut}$ the machine morphism
$M(A)$ defined in (4.3.3) or equivalently in (4.4.3).
For morphisms we have: $M(f:A \to A') = f:S \to S'$.
Vice versa given a $\underline{K\text{-}Medv}$-morphism $g:S \to <I^+,O>$ from (S,d)
to $(<I^+,O>,L)$ the automaton $A = M^{-1}(g)$ is defined by
$A = (S,d,l)$ with $l = ev_1 \circ (g \otimes I)$ (cf. (4.4.3)).

<u>Proof</u>: Given an automaton $A = (S,d,l)$ the machine morphism
$M(A):S \to <I^+,O>$ is a $\underline{K\text{-}Medv}$-morphism equal to g defined
by (4.4.3). Thus we have $l = ev_1 \circ (M(A) \otimes I)$ showing
$M^{-1}(M(A)) = A$ by definition of M^{-1}. On the other hand we
have $M(M^{-1}(g)) = g$ by the uniqueness of g in (4.4.3).

In order to show that M is an isomorphism of categories
with inverse functor M^{-1} it remains to show that for each
pair of automata $A = (S,d,l)$ and $A' = (S',d',l')$ and each
$\underline{K\text{-}Medv}$-morphism $f:S \to S'$ the following conditions are
equivalent:

  (i) $l' \circ (f \otimes I) = l$ and

  (ii) $M(A') \circ f = M(A)$.

Now by (4.3.1) and (4.4.3) (ii) is equivalent to

  (iii) $ev_1 \circ (M(A') \otimes I) \circ (f \otimes I) = ev_1 \circ (M(A) \otimes I)$

and hence by definition of l' and l to (i) .   ∎

Remark: The category $(\underline{K}\text{-}\underline{Medv} \downarrow <I^+,O>)$ is a special case
of a "comma" category defined in 12.8 and theorem 4.5 is a
corollary of lemma 4.4 and a categorical lemma which asserts
that each pair of adjoint functors implies an isomorphism
of the corresponding comma categories (cf. [28,59]).

Corollary (C h a r a c t e r i z a t i o n   o f   M a -
c h i n e   M o r p h i s m s) : A $\underline{K}$-morphism $g:S \to <I^+,O>$
is the machine morphism of an automaton iff there is a
$d:S\otimes I \to S$ such that $g\circ d = L\circ(g\otimes I)$. In this case we have
$g = M(A)$ for the automaton $A = (S,d,l)$ with $l = ev_1\circ(g\otimes I)$.

Proof: Since the condition $g\circ d = L\circ(g\otimes I)$ asserts that g
is a $\underline{K}$-$\underline{Medv}$-morphism the corollary is a direct consequence
of theorem 4.5.   ∎

4.6 Behavior and $\mathfrak{C}$-$\mathfrak{M}$-Factorization: Now we are going to
construct the input-output behavior $E(A)$ of an automaton A.
In 2.1 $E(A)$ was defined as being the image of the machine
function $M(A):S \to <I^+,O>$ consisting of all input-output
functions $M(A)(s) = l^+(s,-):I^+ \to O$ for $s\in S$. In order to
generalize this construction we only need a corresponding
image factorization in our category $\underline{K}$ such that each mor-
phism $f:K \to L$ in $\underline{K}$ has a representation $f = (K \overset{e}{\to} B \overset{m}{\to} L)$
where e belongs to a class $\mathfrak{C}$ of "surjective" morphisms
and m to a class $\mathfrak{M}$ of "injective" morphisms. Moreover,
we will assume that the object B in this representation is
unique up to isomorphism. Then B together with the morphism $m:B \to L$ will be called image of f .

As motivated in 3.2 we take for $\mathfrak{C}$ a class of epimorphisms
and for $\mathfrak{M}$ a class of monomorphisms in $\underline{K}$ (cf. 12.3).
$\mathfrak{C}$-$\mathfrak{M}$-factorizations are studied in [50] for example.

Definition: Let $\mathfrak{C}$ be a class of epimorphisms in $\underline{K}$ including all isomorphisms which is closed under composition,
and $\mathfrak{M}$ a class of monomorphisms in $\underline{K}$ with the same prop-
erty. Then $\underline{K}$ is called to have an $\underline{\mathfrak{C}\text{-}\mathfrak{M}\text{-factorization}}$ if for

72

each morphism f:K → L in <u>K</u> there is a representation
f = (K $\overset{e}{\to}$ B $\overset{m}{\to}$ L) with e∈𝕮 and m∈𝔐 and for each other representation f = (K $\overset{e'}{\to}$ B' $\overset{m'}{\to}$ L) with e'∈𝕮 and m'∈𝔐 there
is a unique isomorphism i:B $\overset{\sim}{\to}$ B' in (4.6.1) such that
i∘e = e' and m'∘i = m .

(4.6.1)

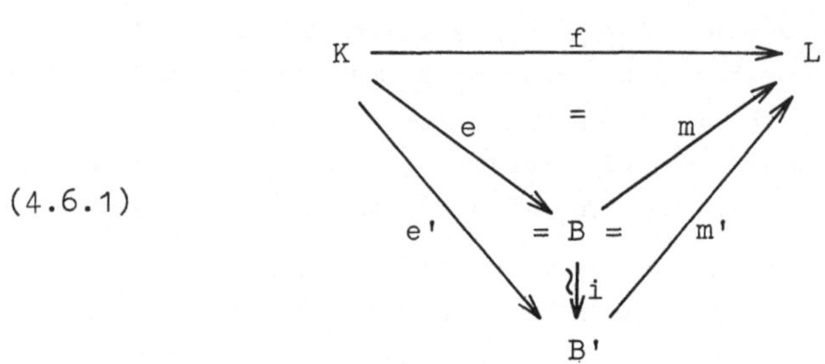

Two morphisms m:B → L in 𝔐 and m':B' → L in 𝔐 are
called <u>equivalent</u> if there is an isomorphism i:B $\overset{\sim}{\to}$ B' such
that m'∘i = m . Thus we get equivalence classes of 𝔐-mor-
phisms with fixed codomain   L and we assume that there is
a canonical representative m:B → L of each equivalence
class. Now an 𝕮-𝔐-factorization f = m∘e will be called
<u>canonical</u> if m is a canonical representative. In this case
m will be called <u>image</u> of f , written m = im(f) .

<u>Examples</u>: 1. In <u>Set</u> we have an 𝕮-𝔐-factorization taking 𝕮
resp. 𝔐 to be the class of all surjective resp. injective
functions which has already been used in 2.3 and 2.2. Moreover we have for each function f:K → L a canonical factorization f = i∘f' where f':K → f[K] is the restriction of
f to the image f[K] = {f(a)∈L/ a∈K} and i:f[K] → L is
the inclusion.

2. In <u>PD</u> we take for 𝕮 the class of all partially defined
surjective functions and for 𝔐 , as before, all injective
functions. Thus we get a canonical 𝕮-𝔐-factorization of each
partially defined function f:K → L taking the restriction
f' of f to the image f[K] followed by the inclusion.

3. In $\underline{\text{Mod}}_R$ $\mathfrak{E}$ resp. $\mathfrak{M}$ is the class of all surjective resp. injective R-linear functions. Thus we get a canonical $\mathfrak{E}$-$\mathfrak{M}$-factorization in analogy to 1.

4. In $\underline{\text{Top}}$ there are several different $\mathfrak{E}$-$\mathfrak{M}$-factorizations. Mostly we will take $\mathfrak{E}$ to be the class of all epimorphisms, i.e. surjective continuous functions, and $\mathfrak{M}$ to be the class of all extremal monomorphisms, i.e. injective continuous functions with subspace topology. But sometimes it is better to take $\mathfrak{E}$ to be the class of all extremal epimorphisms, i.e. surjective functions in $\underline{\text{Top}}$ with identification topology on the codomain, and $\mathfrak{M}$ to be the class of all injective functions in $\underline{\text{Top}}$. Both cases lead to an $\mathfrak{E}$-$\mathfrak{M}$-factorization in $\underline{\text{Top}}$ and in the first case we get a canonical factorization by taking the image of a function with subspace topology.

<u>General Assumptions</u>: Henceforth we will assume that $(\underline{K},\otimes)$ has an $\mathfrak{E}$-$\mathfrak{M}$-factorization which is compatible with $\otimes$ in the sense that for all $e\in\mathfrak{E}$ and objects $K\in\underline{K}$, $e\otimes K$ and $K\otimes e$ are in $\mathfrak{E}$. Moreover we assume to have canonical representatives in $\mathfrak{M}$ and hence a canonical $\mathfrak{E}$-$\mathfrak{M}$-factorization.

<u>Construction and Definition</u>: Given an automaton A and a canonical $\mathfrak{E}$-$\mathfrak{M}$-factorization (4.6.2)

(4.6.2)

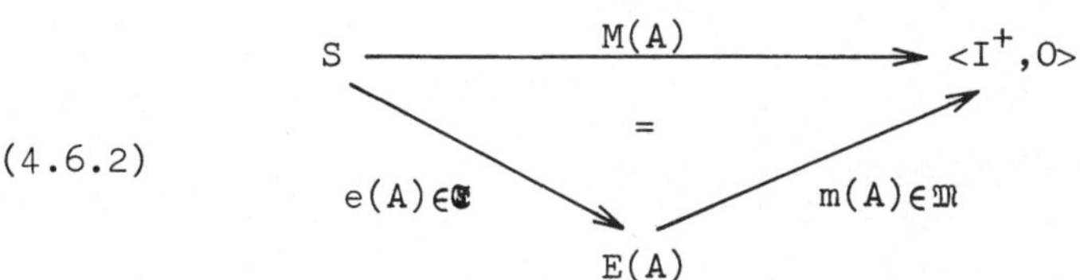

of the machine morphism M(A) of A the image $m(A):E(A) \to \langle I^+,O\rangle$ of M(A) , or short E(A) , is called <u>behavior</u> of A .

In the examples of deterministic, partial and bilinear
automata, E(A) is exactly the subset resp. submodule of all
input-output functions $M(A)(s) = 1^+(s,-):I^+ \to 0$ with $s \in S$.
In the topological case this subset has subspace topology
of $<I^+,O>$ if $\mathfrak{M}$ is the class of extremal monomorphisms
or quotient topology of S if $\mathfrak{E}$ is the class of extremal
epimorphisms.

Our next problem is to characterize those $\mathfrak{M}$-subobjects B
of $<I^+,O>$ which are behaviors of automata (cf. 2.5).
Therefore we need the following "diagonal lemma" (cf. [50]),
called "Zeiger-fill-in-lemma" in [56], for $\mathfrak{E}$-$\mathfrak{M}$-factorizations
and the compatibility with $\underline{K}$-$\underline{Medv}$-morphisms (cf. (2.3.3)).

<u>4.7 Lemma</u> ($\mathfrak{E}$ - $\mathfrak{M}$ - F a c t o r i z a t i o n) : 1. Given
morphisms e, f, g and m in $\underline{K}$ with $g \circ e = m \circ f$ and $e \in \mathfrak{E}$,
$m \in \mathfrak{M}$ there is a unique "diagonal morphism" d such that the
following diagram (4.7.1) is commutative:

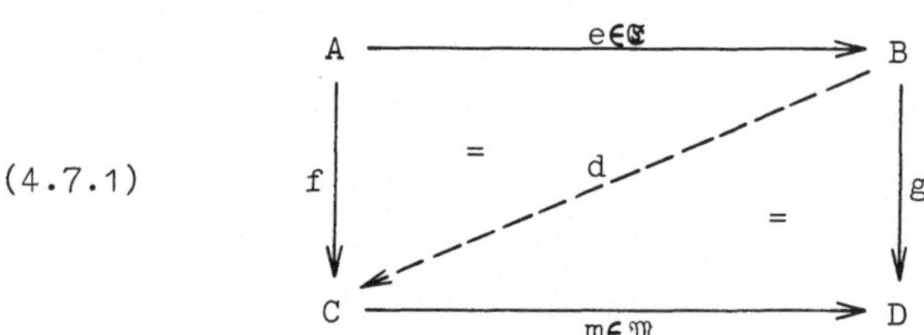

(4.7.1)

2. Given a $\underline{K}$-$\underline{Medv}$-morphism $f:(S,d) \to (S',d')$ and an $\mathfrak{E}$-$\mathfrak{M}$-
factorization $f = (S \xrightarrow{e} \bar{S} \xrightarrow{m} S')$ of the $\underline{K}$-morphism $f:S \to S'$
then there is a unique $\bar{d}:\bar{S} \otimes I \to \bar{S}$ such that $e:(S,d) \to (\bar{S},\bar{d})$
and $m:(\bar{S},\bar{d}) \to (S',d')$ are $\underline{K}$-$\underline{Medv}$-morphisms.

3. Given $g \circ f \in \mathfrak{E}$ we have $g \in \mathfrak{E}$ such that the class $\mathfrak{E}$ satisfies the assumptions in 3.2. Dually $g \circ f \in \mathfrak{M}$ implies $f \in \mathfrak{M}$.

4. Given morphisms $r:A \to B$ and $c:B \to A$ with $r \circ c = id_B$
we have $r \in \mathfrak{E}$ and $c \in \mathfrak{M}$. Thus the class of all retractions

and the class of all coretractions (cf. 12.3) is contained in $\mathfrak{E}$ and $\mathfrak{M}$ respectively. Moreover the class of all isomorphisms is exactly the intersection of $\mathfrak{E}$ and $\mathfrak{M}$.

<u>Proof</u>: 1. Let $f = (A \xrightarrow{e_1} X \xrightarrow{m_1} C)$ and $g = (B \xrightarrow{e_2} Y \xrightarrow{m_2} D)$ be $\mathfrak{E}$-$\mathfrak{M}$-factorizations of f and g respectively. Then $m_2 \circ (e_2 \circ e) = g \circ e$ and $(m \circ m_1) \circ e_1 = m \circ f$ are two $\mathfrak{E}$-$\mathfrak{M}$-factorizations of $g \circ e = m \circ f$. Hence there is a unique isomorphism $i : Y \xrightarrow{\sim} X$ such that diagrams (1) and (2) in (4.7.2) are commutative. Taking $d := m_1 \circ i \circ e_2$ we get the desired diagonal morphism. Uniqueness of d follows from the fact that e is an epimorphism with $d \circ e = f$ and hence right cancellable.

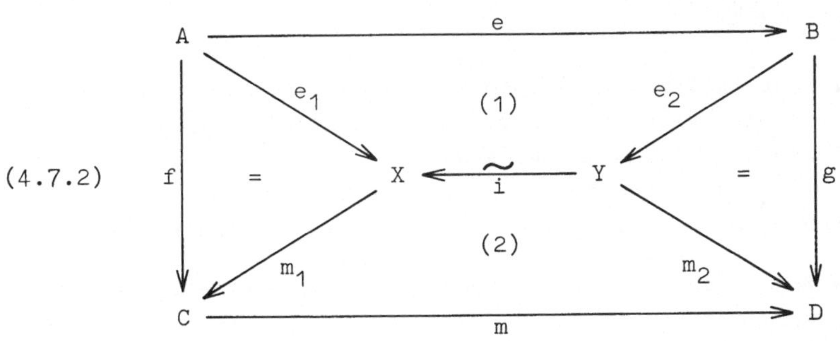

2. The existence and uniqueness of $\bar{d} : \bar{S} \otimes I \to \bar{S}$ is given by the first part of the lemma applied to the following commutative diagram (4.7.3) with $e \otimes I \in \mathfrak{E}$ by assumption on $\mathfrak{E}$ and $m \in \mathfrak{M}$ and the definition of <u>K-Medv</u>-morphisms.

3. Let $m \circ e = g$ be an $\mathfrak{E}$-$\mathfrak{M}$-factorization of g . Since $g \circ f \in \mathfrak{E}$ and $m \in \mathfrak{M}$ there is a diagonal morphism d satisfying $d \circ (g \circ f) = e \circ f$ and $m \circ d = id$. Thus we have $m \circ d \circ m = m$ and hence $d \circ m = id$ because m is a monomorphism showing that m is an isomorphism. Thus $g = m \circ e$ belongs to $\mathfrak{E}$ because $\mathfrak{E}$ is closed under the composition with isomorphisms (cf. 4.6). The second part is dual.

4. Since id_B belongs to $\mathfrak{E}$ and $\mathfrak{M}$ and $r \circ c = id_B$ we have

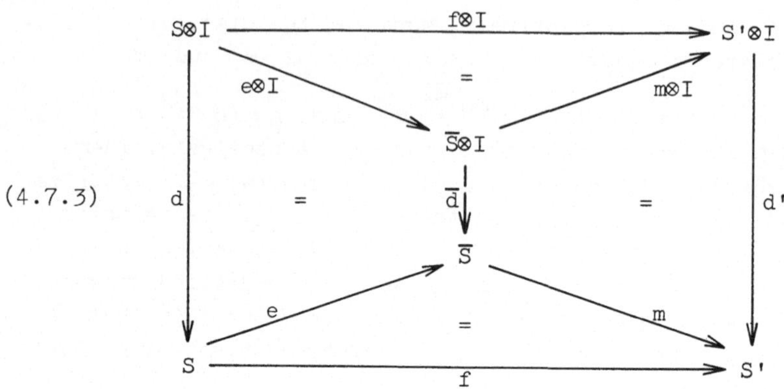

re$\mathfrak{C}$ and   c$\in\mathfrak{M}$ by 3. Especially each isomorphism belongs to
$\mathfrak{C}$ and $\mathfrak{M}$. Vice versa given a morphism $f:A \to B$ which
belongs to $\mathfrak{C}$ and $\mathfrak{M}$ we have two $\mathfrak{C}$-$\mathfrak{M}$-factorizations
$f = f \circ id_A$ and $f = id_B \circ f$ of f which are unique up to isomorphism by definition. Hence f is an isomorphism.   ∎

Now we are able to give a behavior characterization which is
a generalization of theorem 2.5 for deterministic automata.

<u>4.8 Theorem</u> (B e h a v i o r   C h a r a c t e r i z a -
t i o n) :   A <u>K</u>-morphism $m:B \to \langle I^+,O\rangle$ is the behavior of
an automaton A iff m is a canonical representative of a
morphism in $\mathfrak{M}$ (cf. 4.6) and is closed under left shift L
(cf. 4.4) meaning that there is a morphism $L':B\otimes I \to B$ in
(4.8.1) satisfying $m \circ L' = L \circ (m \otimes I)$ such that m becomes a
<u>K-Medv</u>-morphism.

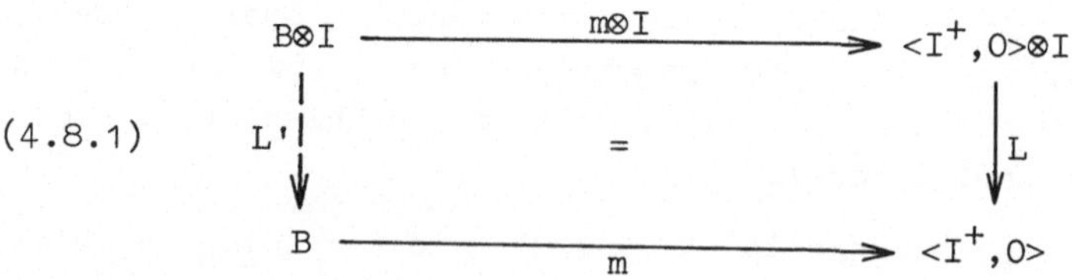

In this case the automaton $A = (B,L',1)$ with $1 := ev_1 \circ (m \otimes I)$ is a realization of the behavior m , i.e. $m(A) = m$.

<u>Interpretation</u>: Regarding our examples in 4.6 the canonical representatives are given by subsets $B \subseteq \;<I^+,O>$ and all those subsets B are realizable which are closed under left shift, i.e. for each $f \in B$ and $x \in I$ we have $f \circ L_x \in B$ (cf.2.5). Of course in the case $\underline{Mod}_R$ each B is assumed to be a sub-module of $<I^+,O>$ and a subspace in the case $\underline{Top}$ with $\mathfrak{M}$ being the class of extremal monomorphisms.

<u>Proof</u>: Given an automaton A the behavior of A is the image $m(A):E(A) \to \;<I^+,O>$ of the machine morphism $M(A) = m(A) \circ e(A)$ which is a $\underline{K\text{-Medv}}$-morphism by theorem 4.5. Hence by lemma 4.7,2 there is a $\bar{d}:E(A) \otimes I \to E(A)$ such that $m(A)$ satisfies $m(A) \circ \bar{d} = L \circ (m(A) \otimes I)$. Moreover $m(A)$ is a canonical representative by construction. Vice versa given $m:B \to \;<I^+,O>$ satisfying the above conditions, m is the machine morphism of the automaton $A = (B,L',1)$ with $1 = ev_1 \circ (m \otimes I)$ by the corollary of theorem 4.5. Thus we have $m = M(A) = m(A)$ because m is already a canonical represent-ative.　　　　　　　　　　　　　　　　　　　　　　　　■

<u>Remark</u>: In order to characterize the behaviors of "finite" automata we assume to have a class $\underline{F}$ of "finite" objects in $\underline{K}$ such that for each morphism $f:F \to K$ in $\mathfrak{S}$ we have: $F \in \underline{F}$ implies $K \in \underline{F}$. Take for example finite sets in $\underline{Set}$ or $\underline{PD}$, finitely generated R-modules in $\underline{Mod}_R$ or (quasi-)compact spaces in $\underline{Top}$. These properties are preserved by morphisms in $\mathfrak{S}$ which are surjective in all our examples.

Since $e(A):S \to E(A)$ lies in $\mathfrak{S}$ we have the following corollary of theorem 4.8 :

<u>Corollary</u> (F i n i t e   B e h a v i o r s) : A behavior $m:B \to \;<I^+,O>$ is realizable by an automaton $A = (S,d,1)$ with "finite" state object $S \in \underline{F}$ iff B is "finite", i.e. $B \in \underline{F}$.

78

<u>4.9 Examples of Automata in Closed Categories</u>: Motivated by
the example of deterministic automata we have given the
basic notions for a theory of automata in closed categories.
The general assumptions for the monoidal category $(\underline{K},\otimes)$
given in 4.1 have been verified for the categories $(\underline{Set},\times)$,
$(\underline{PD},\times)$, $(\underline{Mod}_R,\otimes)$ and $(\underline{Top},\otimes)$ in 4.2, 4.3 and 4.6.
Thus we have developed a common theory for deterministic,
partial, bilinear and topological automata (cf. 1.12).
Moreover there are several other examples of closed categories satisfying the general assumptions in 4.1. Thus we get
a large number of automata which are applications of our
theory. Some of them we will give in the following list
with references to chapter 1 or to the literature. Explanations are given below.

1. $(\underline{Set},\times)$ for deterministic automata (cf. 1.1)

2. $(\underline{PD},\times)$ for partial automata (cf. 1.13)

3. $(\underline{Mod}_R,\otimes)$ for bilinear automata (cf. 1.6)

4. $(\underline{Top},\otimes)$ for topological automata (cf. 1.7)

5. $(\underline{CG},\pi)$ for compactly generated topological automata
                 (cf. [44,28,32], for other topological exam-
                 ples of closed categories cf. [21,74,77])

6. $(\underline{Metr},\times)$ for metric automata (cf. [5,32])

7. $(\underline{Tol},\otimes)$ for tolerance automata (cf. [32,78])

8. $(\underline{LTop},\otimes)$ for linear topological automata (cf. [73])

9. $(\underline{PDTop},\otimes)$ for partial topological automata (cf. [32])

10. $(\underline{Cat},\times)$ for category automata (cf. [11])

11. $(\underline{Aff},\otimes)$ for affine automata (cf. [44]) .

Finally there are some examples of monoidal categories which
do not satisfy all the assumptions in 4.1 but can be treated
in a similar way because they are of deterministic type:

12. $(\underline{Mod}_R,\times)$ for linear automata (cf. 1.6)

13. $(\underline{Top},\times)$ for topological automata (cf. 1.7, [32])

14. ($\underline{\text{Graph}}_M$,⊗) for graph automata (cf. [28]).

<u>Explanations</u>: 1.-4. The categories ($\underline{\text{Set}}$,×) of sets with
cartesian product, ($\underline{\text{PD}}$,×) of partially defined functions
with cartesian product, ($\underline{\text{Mod}}_R$,⊗) of R-modules with tensorproduct and ($\underline{\text{Top}}$,⊗) of topological spaces with biproduct
are discussed in 1.10, 1.13, 4.2, 4.3 and 4.6.

5. ($\underline{\text{CG}}$,π) is the category of compactly generated Hausdorff-
spaces and π the Kelleyfication of the topological product.
($\underline{\text{CG}}$,π) is closed taking the Kelleyfication of the compact
open topology on the function space <K,O> .

6. The category of metric spaces ($\underline{\text{Metr}}$,×) with distance
$d \leq 1$ and distance decreasing (or better not increasing) mappings is closed defining the tensorproduct $(A,d) \times (A',d')$
to be $(A \times A', \min(d+d',1))$ and <(A,d),(A',d')> to be the
set of all decreasing mappings $f: A \to A'$ with metric defined by $d(f,g) := \sup\{d'(f(a),g(a))/\ a \in A\}$.

7. Tolerance spaces are pairs (A,r) of a set A together
with a reflexive symmetric relation r and tolerance morphisms $f: A \to A'$ are functions such that x r y implies
$f(x)\, r'\, f(y)$. The category ($\underline{\text{Tol}}$,⊗) is again closed taking
$(A,r) \otimes (A',r')$ to be $(A \times A', r \times r')$ and <(A,r),(A',r')> to
be the set of all tolerance morphisms $f: A \to A'$ with
$f\, r_{<,>}\, g$ iff for all $x,y \in A$ x r y implies $f(x)\, r'\, g(y)$.

8. There are several categories of linear topological spaces
satisfying our assumptions, e.g. seminormed, bornological or
locally convex spaces (cf. [73]).

9. Topological spaces together with partially defined continuous functions and the biproduct ⊗ lead to a closed
category ($\underline{\text{PDTop}}$,⊗) where <K,O> is the space of all non-
empty partially defined continuous functions with topology
of pointwise convergence (cf. examples 2 and 4 in 4.3).

10. The category ($\underline{\text{Cat}}$,×) of all small categories and functors together with the product of categories × is closed.
In [11] automata based on ($\underline{\text{Cat}}$,×) are interpreted as

parsers for formal languages.

11. In [44] the closed category $(\underline{Aff},\otimes)$ of R-modules and affine functions, i.e. the composition of linear functions and translations, is proposed to become very important in linear system theory.

12. The category $(\underline{Mod}_R,\times)$ of R-modules with direct product is not closed because the direct product does not preserve coproducts, which are direct sums in this case, and this would be a necessary condition for $(\underline{Mod}_R,\times)$ to be closed (cf. 4.3 assumption). However, using the theory of systematics, sketched in chapter 3, it is shown in [28] that the systematic of linear automata satisfies the Minimal Realization Principle and thus we get nearly the same results concerning reduction, minimization and realization as in the deterministic case.

13. Again the category $(\underline{Top},\times)$ of topological spaces with topological product is not closed because the functor $-\times I$ does not have a right adjoint in general. But if I is locally compact we have a right adjoint. In fact restricting the input object I to be locally compact the whole theory for automata in closed categories remains true for this type of topological automata.

14. Let $\underline{Graph}_M$ be the category of all graphs with fixed set M of vertices and all graph morphisms which are identical on M. $G\otimes G'$ is defined to be the graph which has as edges from m to n all pairs (e,e') of edges e in G from m to p and e' in G' from p to n for arbitrary p with $m,n,p \in M$. It is easy to see that $\otimes$ is not symmetric but all the other assumptions given in 4.1 are valid for $(\underline{Graph}_M,\otimes)$ and the most part of our theory remains valid for graph automata. More details are given in [28,59].

<u>4.10 Remark</u> (G e n e r a l i z a t i o n s) :
1. Due to our general convention 1.11 we have fixed the

input object I and output object O which is convenient
for our notation and sufficient for our applications.
However it is possible to generalize most of our results to
variable I and O taking arbitrary $\underline{K}$-morphisms $f_O:O \to O'$
but only $\underline{K}$-retractions $f_I:I \to I'$ in the O- and I-component
respectively. Especially for theorem 4.5 we need a retrac-
tion in the I-component.

2. As remarked in 1.14 and 4.4 our theory can be reformulated for Moore-automata replacing the output morphism
$l:S \otimes I \to O$ by $m:S \to O$ and the machine morphism has the
form $M(A):S \to \,<I^*,O>$ instead of $M(A):S \to \,<I^+,O>$.
An interesting generalization of Moore-automata in monoidal
closed categories is the approach of adjoint machines in
[5] (cf. 4.4).

5. Reduction and Minimization of Automata in Closed Categories

The basic concepts for automata in closed categories have
already been studied in chapter 4. Now we want to show that
the behavior construction can be extended to a behavior functor which satisfies the Minimal Realization Principle stated
in 3.6. By theorem 3.6 we get a systematic of automata in
closed categories which has a minimal, realizing and reduced
subsystematic. Thus it is possible to apply all the general
results of chapter 3 to automata in closed categories.
Especially we now get all the results concerning reduction,
minimization and realization,which were sketched for the
deterministic case in chapter 2,now for automata in arbitrary closed categories satisfying the general assumptions
in 4.1. For examples we refer to the list in 4.9.

$\underline{5.1\ General\ Assumptions}$: According to 4.1 we assume that the
monoidal category $(\underline{K},\otimes)$ is closed, has countable coproducts
and a canonical $\mathfrak{E}$-$\mathfrak{M}$-factorization. $\underline{K}$-$\underline{Aut}$ is the category
of automata in $(\underline{K},\otimes)$ which will be extended in 5.2 to the
systematic $\underline{\underline{K}}$-$\underline{\underline{Aut}} = (\underline{K}$-$\underline{Aut},\underline{B},E)$ of automata in $(\underline{K},\otimes)$.

Motivated by the behavior characterization in theorem 4.8
and the remarks in 3.1 we start with the construction of a
behavior category $\underline{B}$ and a behavior functor $E:\underline{K}$-$\underline{Aut} \to \underline{B}$
leading to the systematic $\underline{\underline{K}}$-$\underline{\underline{Aut}}$ of automata in $(\underline{K},\otimes)$.

$\underline{5.2\ Construction\ and\ Definition}$ (B e h a v i o r   F u n c -
t o r   a n d   S y s t e m a t i c   o f   A u t o m a t a
i n   C l o s e d   C a t e g o r i e s) :

1. The $\underline{behavior\ category}$ $\underline{B}$ is defined as being a full subcategory of the category $(\underline{K}$-$\underline{Medv} \downarrow <I^+,O>)$ defined in 4.5,
such that $\underline{B}$ has as its objects one canonical representative

for each class of those equivalent $\underline{K}$-$\underline{Medv}$-morphisms
$m:B \rightarrow <I^+,O>$ which are in $\mathfrak{M}$ (cf. (4.6.1)). Hence there is
at most one morphism between each pair of objects in $\underline{B}$
such that $\underline{B}$ is a partially ordered class.

2. The behavior construction $E(A)$ of an automaton $A \in \underline{K}$-$\underline{Aut}$
can be extended to a <u>behavior functor</u>

$$E:\underline{K}\text{-}\underline{Aut} \rightarrow \underline{B} .$$

$E(A)$ is defined as being the image of the machine morphism
$M(A)$ (cf. 4.6), and for $f:A \rightarrow A'$ in $\underline{K}$-$\underline{Aut}$ $E(f):E(A) \rightarrow$
$\rightarrow E(A')$ is the unique diagonal morphism in the following
diagram (5.2.1) (cf. 4.7 and remark below).

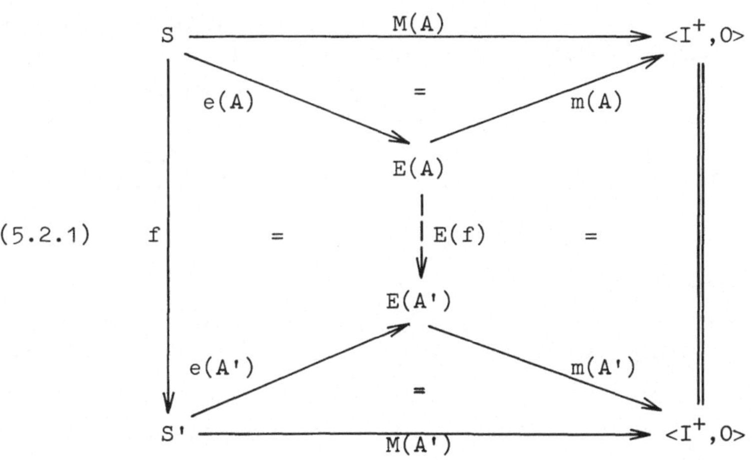

3. $\underline{\underline{K}}$-$\underline{\underline{Aut}} = (\underline{K}$-$\underline{Aut},\underline{B},E)$ is called <u>systematic of automata</u> in
$(\underline{K},\otimes)$ (cf. 3.1).

<u>Remarks</u>: In the examples of deterministic, partial, bilinear
and topological automata the behavior category is the set $\underline{B}$
of all those subsets resp. subobjects $B \subseteq <I^+,O>$ which are
closed under left shift (cf. 4.8). The set $\underline{B}$ is partially
ordered by inclusion where the inclusions are in fact $\underline{K}$-$\underline{Medv}$-

morphisms. It remains to prove some details concerning the
behavior functor:

First of all the behavior $E(A)$, or more precisely
$m(A):E(A) \to <I^+,O>$, is an object in $\underline{B}$ by theorem 4.8.
Moreover given a morphism $f:A \to A'$ in $\underline{K}$-$\underline{Aut}$,
$M(f) = f:S \to S'$ is a morphism from $M(A)$ to $M(A')$ by the-
orem 4.5, i.e. $M(A') \circ f = M(A)$. Since $e(A) \in \mathfrak{C}$ and $m(A') \in \mathfrak{M}$
we can apply lemma 4.7,1 to the diagram (5.2.1) yielding a
unique $\underline{K}$-morphism $E(f):E(A) \to E(A')$ such that $E(f) \circ e(A) =$
$= e(A') \circ f$ and $m(A') \circ E(f) = m(A)$. Since $m(A)$ and $m(A')$
are in $\underline{K}$-$\underline{Medv}$ by lemma 4.7,2 it is easy to check that $E(f)$
is in $\underline{K}$-$\underline{Medv}$, too (using the fact that $m(A')$ is a monomorphism) and hence a morphism in $\underline{B}$ from $E(A)$ to $E(A')$.
The remaining functor properties of E are obvious, since $\underline{B}$
is a partially ordered class. In fact, if there are two morphisms f_1 and f_2 from m to m' in $\underline{B}$ we have
$m' \circ f_1 = m = m' \circ f_2$ and hence $f_1 = f_2$ since m' is a monomorphism. Moreover by construction there are no different iso-
morphic objects in $\underline{B}$. Hence $\underline{B}$ is a partially ordered
class.

<u>5.3 Theorem</u> (M i n i m a l   R e a l i z a t i o n) :
The systematic $\underline{\underline{K}\text{-}\underline{\underline{Aut}}} = (\underline{K}\text{-}\underline{Aut},\underline{B},E)$ of automata in the closed
category $(\underline{K},\otimes)$ satisfies the Minimal Realization Principle
defined in 3.6.

More specifically we have a minimal realization functor
$M^*:\underline{B} \to \underline{K}\text{-}\underline{Aut}$ defined by

$$M^* := (\underline{B} \xrightarrow{\ J\ } (\underline{K}\text{-}\underline{Medv} \downarrow <I^+,O>) \xrightarrow{\ M^{-1}\ } \underline{K}\text{-}\underline{Aut})$$

where J is the inclusion functor and M^{-1} the inverse
functor of M given in 4.5. Moreover M^* satisfies
$E \circ M^* = Id_{\underline{B}}$ and for all automata A in $\underline{K}$-$\underline{Aut}$ and all behaviors $m:\overline{B} \to <I^+,O>$ in $\underline{B}$ satisfying $E(A) \subseteq B$ there is a
unique $\underline{K}$-$\underline{Aut}$-morphism $f:A \to M^*(m)$ which belongs to the
class $\mathfrak{C}$ in the case $E(A) = B$. The minimal realization
$M^*(m)$ is explicitly given by

$$M*(m) = A := (B,L',l)$$

where $L':B\otimes I \to B$ is the restriction of L (cf. (4.8.1))
and $l := ev_1 \circ (m\otimes I)$ (cf. (4.4.2)).
Finally minimal realization is unique up to isomorphism
meaning that each minimal automaton A with behavior
$m(A) = m$ is isomorphic to the minimal realization $M*(m)$.

<u>Proof</u>: Given $m:B \to <I^+,O>$ in <u>B</u> we have
$E(M*(m)) = E(M^{-1}J(m)) = E(M^{-1}(m)) = m$ since $E(M^{-1}(m))$ is the
image of the machine morphism $M(M^{-1}(m)) = m$ and hence equal
to m . Thus we have $E \circ M* = Id_{\underline{B}}$ because the equality on
morphisms follows from the fact that <u>B</u> is partially or-
dered. Now given $A \in \underline{K}\text{-}\underline{Aut}$ with $E(A) \subseteq B$, which means that
there is a <u>K</u>-<u>Medv</u>-morphism $i:E(A) \to B$ satisfying $m \circ i = m(A)$,
we define $f := i \circ e(A)$ in (5.3.1) where $m(A) \circ e(A) = M(A)$
is the canonical $\mathfrak{E}$-$\mathfrak{M}$-factorization of $M(A)$ (cf. 4.6).

(5.3.1)

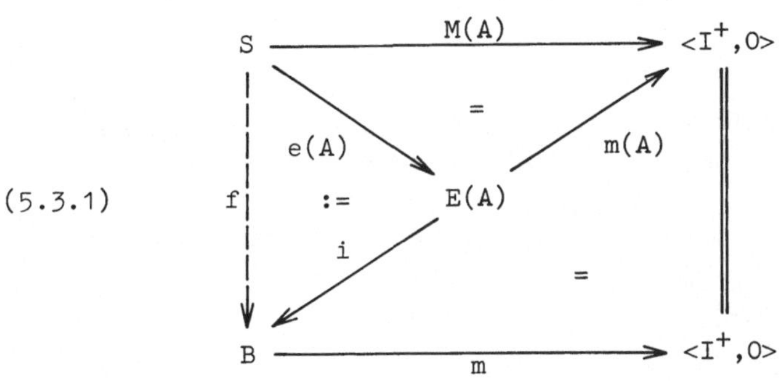

Now $f = i \circ e(A)$ is a morphism in the category
$(\underline{K}\text{-}\underline{Medv} \downarrow <I^+,O>)$ (cf. 4.5) because $e(A)$ and i and
hence $i \circ e(A)$ are <u>K</u>-<u>Medv</u>-morphisms satisfying
$m \circ f = m \circ i \circ e(A) = m(A) \circ e(A) = M(A)$. Moreover f is unique,
because $m \in \mathfrak{M}$ is a monomorphism, and hence a unique morphism
$f:A \to M*(m)$ in <u>K</u>-<u>Aut</u> using the isomorphism M^{-1} of M
in (4.5) and the identities $M^{-1}(M(A)) = A$ as well as
$M^{-1}(m) = M*(m)$. Finally in the case $E(A) = B$ implying

$i = id_{E(A)}$ we have $f = i \circ e(A) = e(A) \in \mathfrak{E}$. In 5.4 we will
show that in this case   f   is already a "reduction".
The explicit representation of M*(m)   follows from the definition of M^{-1} in theorem 4.5. The uniqueness of M*(m)
is a consequence of 3.6,5.   ∎

Remark: The fact that K̲-A̲u̲t̲ satisfies the Minimal Realization Principle is very important because the problems of
reduction and minimization are already solved in such a
situation. Using theorem 3.6 we know that there is a unique
minimal, reduced and realizing subsystematic S̲' of K̲-A̲u̲t̲
which will be shown to be equal to the subsystematic of all
"observable" or "𝔐-minimal" automata. According to 2.4   A
will be called observable if the machine morphism M(A) of
A belongs to the class 𝔐 . Moreover the theorem in 2.4
becomes a corollary of our result.

First we have to generalize the definitions in 2.4:

5.4 Definitions:

1. An automaton A is called observable (or 𝔐-minimal) if
   the machine morphism M(A) belongs to the class 𝔐 of
   monomorphisms. Let K̲-A̲u̲t̲$_\mathfrak{M}$ be the full subcategory of
   K̲-A̲u̲t̲ consisting of all observable automata then
   K̲-A̲u̲t̲$_\mathfrak{M}$ = (K̲-A̲u̲t̲$_\mathfrak{M}$,B̲,E') is a subsystematic of K̲-A̲u̲t̲ -
   = (K̲-A̲u̲t̲,B̲,E) where E' is the restriction of E to
   K̲-A̲u̲t̲$_\mathfrak{M}$.

2. According to 3.1 and 3.2 automata A and A' are called
   equivalent if E(A) = E(A'), an automata morphism
   f:A → A' in K̲-A̲u̲t̲ is called reduction if the K̲-mor-
   phism f:S → S' belongs to the class 𝔈 of epimorphisms.
   Finally A is called reduced if each reduction f:A → A'
   is already an isomorphism of automata.

Remark: In 3.2 we have assumed that a reduction f:A → A'
satisfies E(A) = E(A'). This property is a corollary in our
case because we have (e(A')∘f)∈𝔈 such that m(A')∘(e(A')∘f)

and m(A)∘e(A) are two canonical $\mathfrak{E}$-$\mathfrak{M}$-factorizations of M(A)
(cf. (5.2.1)). Thus E(f):E(A) → E(A') is the identity,
since canonical $\mathfrak{E}$-$\mathfrak{M}$-factorizations are unique (cf. 4.6).

5.5 Theorem (R e d u c t i o n   a n d   M i n i m i z a -
t i o n) : The systematic $\underline{\underline{K}}$-$\underline{\underline{Aut}}_{\mathfrak{m}}$ of all observable automata is a minimal, reduced and realizing subsystematic of
$\underline{\underline{K}}$-$\underline{\underline{Aut}}$.
Hence there are constructions for reduction and minimization
of automata in $(\underline{\underline{K}},\otimes)$ and these constructions are equal and
unique up to isomorphism. In more detail we have:

1. (M i n i m i z a t i o n) : For each automaton A in
$\underline{\underline{K}}$-$\underline{\underline{Aut}}$ there is an equivalent observable automaton $\bar{M}(A)$
which is minimal in the following sense: For each automaton
A' with $E(A') \subseteq E(\bar{M}(A)) = E(A)$ there is a unique automata
morphism $f:A' \to \bar{M}(A)$, which is a reduction in the case
E(A') = E(A). Moreover equivalent observable automata are
isomorphic and the construction of $\bar{M}(A)$ is defined by
$\bar{M}(A) = M*E(A) = M^{-1}(m(A))$ (cf. 4.5, 4.6, 4.8), where
m(A)∘e(A) = M(A) is the canonical $\mathfrak{E}$-$\mathfrak{M}$-factorization in the
following diagram (5.5.1) :

(5.5.1)

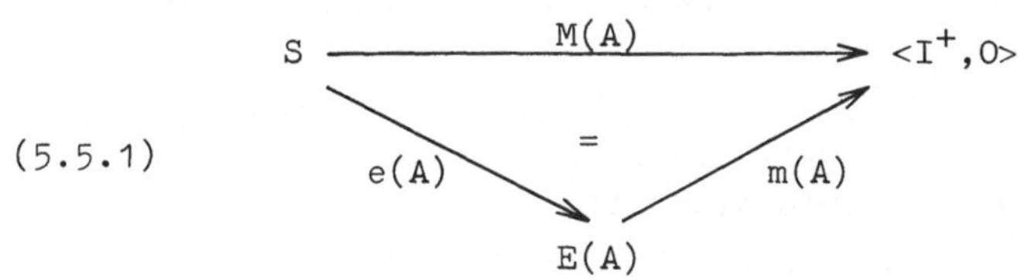

Explicitly $\bar{M}(A)$ is given by $\bar{M}(A) = (E(A),\bar{d},\bar{I})$ where
$\bar{d}:E(A)\otimes I \to E(A)$ is given by lemma 4.7,2 and
$\bar{I} = ev_1 \circ (m(A)\otimes I)$ (cf. (4.4.3)).

2. (R e d u c t i o n) : For each automaton A in $\underline{\underline{K}}$-$\underline{\underline{Aut}}$
there is a reduction e(A):A → R(A) to an equivalent observ-
able automaton R(A) which is reduced, and satisfies the

following universal property: For each automata morphism
$f:A \to A'$ from A to a reduced automaton A' there is a
unique morphism $f':R(A) \to A'$ in (5.5.2) satisfying
$f' \circ e(A) = f$.

(5.5.2)

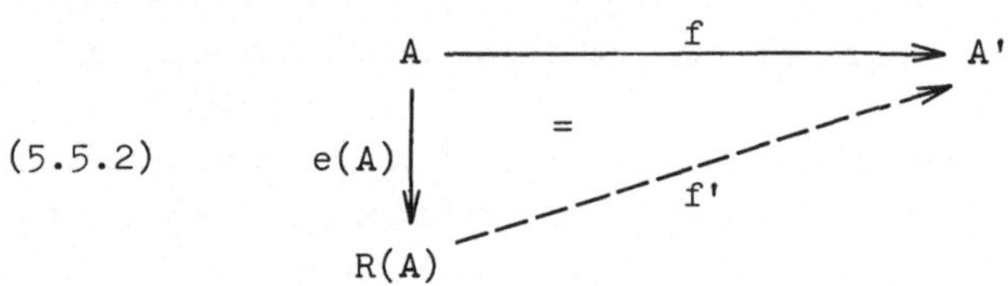

The reduced automaton $R(A)$ can be chosen to be equal to
$\bar{M}(A)$, and $e(A):S \to E(A)$ is an automata morphism
$e(A):A \to R(A)$. The construction of $R(A) = \bar{M}(A)$ can be extended to a reduction functor

$$R: \underline{K\text{-}Aut} \longrightarrow \underline{K\text{-}Aut}_{\mathfrak{M}} \; .$$

Moreover the reduction process is decomposable such that for
each reduction $f:A \to A_1$ there is a unique reduction
$f_1:A_1 \to R(A)$ satisfying $f_1 \circ f = e(A)$.

3. (E q u a l i t y) : Reduction and minimization coincide and they are unique up to isomorphism, and for each
observable automaton A' equivalent to A we have
$A' \cong R(A) = \bar{M}(A)$. Moreover the following conditions are
equivalent for each automaton A' :

   (i) A' is observable

   (ii) A' is reduced

   (iii) A' is minimal (cf. 3.2 and 5.5,1) .

4. (E q u i v a l e n c e) : Given two automata A_1 and
A_2 the following conditions imply each other:

   (i) A_1 and A_2 are equivalent

   (ii) A_1 and A_2 are R-equivalent (cf. 3.4,4)

(iii) $R(A_1)$ and $R(A_2)$ are isomorphic .

5. (F i n i t e   C a r d i n a l i t y) : Given a cardinality function in the sense of 3.7 we have the following characterization: An automaton A' with finite cardinality is minimal (and hence observable by 5.5,3) iff A' has minimal cardinality, i.e. for each equivalent automaton A we have card(A') ≤ card(A) .

Proof: Since K-Aut satisfies the Minimal Realization Principle with minimal realization functor M* (cf. 5.3) the subsystematic S' of K-Aut , defined by all automata which are isomorphic to minimal realizations M*(m) for all m∈B , is minimal, reduced and realizing by theorem 3.6. Thus it suffices to show that S' is equal to K-Aut$_\mathfrak{M}$. All the other properties are corollaries of the theorems in 3.4, 3.6 and 3.7 using the fact that e(A) and m(A) are already K-Medv-morphisms by lemma 4.7,2 such that m(A) is an object and e(A)∈$\mathfrak{E}$ a morphism in the category (K-Medv ↓ <I$^+$,O>) and hence in K-Aut by theorem 4.5. Now we are going to show S' = K-Aut$_\mathfrak{M}$. For each minimal realization M*(m) we have M(M*(m)) = m∈$\mathfrak{M}$ such that M*(m) and thus each isomorphic automaton is observable. Vice versa given an observable automaton A the morphism e(A) in the $\mathfrak{E}$-$\mathfrak{M}$-factorization m(A)∘e(A) = M(A) ∈ $\mathfrak{M}$ is an isomorphism in K using the uniqueness of $\mathfrak{E}$-$\mathfrak{M}$-factorizations. Thus e(A):S $\xrightarrow{\sim}$ E(A) is an isomorphism in (K-Medv ↓ <I$^+$,O>) from M(A) to m(A) and hence A is isomorphic to M*(m(A)) in K-Aut.   ∎

Interpretation: In our examples of deterministic, partial, bilinear and topological automata we get the reduced automaton R(A) which is equal to the minimization $\bar{M}$(A) of A in the following way: Take E(A) as state object, $\bar{d}$:E(A)×I → E(A) as the restriction of the left shift L:<I$^+$,O>×I → <I$^+$,O> and $\bar{I}$:E(A)×I → O defined by $\bar{I}$(f,x) = f(x) for all f∈E(A) iff f(x) is defined and undefined otherwise. In the case of bilinear resp. topological

automata E(A) is a submodule resp. subspace of $\langle I^+,O\rangle$ and
the functions $\bar{d}$ and $\bar{I}$ are bilinear resp. bicontinuous.
The characterization of automata with finite minimal cardinality is useful for finite deterministic, finite partial
and those bilinear automata which have finite dimensional
state vector spaces.

Now we are going to generalize the considerations in 2.8
concerning monoids and monoid automata:

<u>5.6 Monoid Automata and Transition Monoids</u>: As shown in 1.2
and 2.8 the state transition function $d:S\times I \to S$ of a deterministic automaton $A = (I,O,S,d,1)$ can be extended to
$d^*:S\times I^* \to S$ leading to the transition function
$t(A):I^* \to \langle S,S\rangle$ and the transition monoid T(A) of A .
$t(A)$ was defined by $t(A)(w)(s) = d^*(s,w)$ for all $w\in I^*$, $s\in S$,
and T(A) was the image of the function $t(A)$, i.e.
$T(A) = \{d_w:S \to S/\ w\in I^*\}$ with $d_w = d^*(-,w)$.
Using our assumptions in 5.1 these constructions can be generalized to automata in closed categories. Due to 2.8 we
replace the free monoid I^* by an arbitrary monoid M in
$(\underline{K},\otimes)$ and it suffices to consider monoid automata without
output for the moment using the general convention in 1.9:

<u>Definitions</u> (M o n o i d s   a n d   M o n o i d
A u t o m a t a) :

1. A <u>monoid</u> in $(\underline{K},\otimes,U)$ is a 3-tuple $(M,m,1)$, short M, where
   M is an object in $\underline{K}$ and $m:M\otimes M \to M$, $1:U \to M$ are
   $\underline{K}$-morphisms, called multiplication and unit respectively,
   which satisfy the following conditions:
      (i) $m\circ(1\otimes M) = id_M = m\circ(M\otimes 1)$ (unit compatibility)
     (ii) $m\circ(M\otimes m) = m\circ(m\otimes M)$ (associativity) .

2. A <u>monoid automaton</u> in $(\underline{K},\otimes)$ is a 3-tuple $A = (M,S,d)$
   where M is a monoid, S an object and $d:M\otimes S \to S$ a
   morphism in $\underline{K}$ satisfying
            $d\circ(1\otimes S) = id_S$ and
            $d\circ(m\otimes S) = d\circ(M\otimes d)$.

Note that d is now a left action of M on S whereas it was a right action in 2.8. This change has only technical reasons which are used in (5.6.1) and (5.6.2).

3. A <u>monoid morphism</u> $f:M \to M'$ is a <u>K</u>-morphism satisfying
$$f \bullet 1 = 1' \qquad \text{and}$$
$$m' \circ (f \otimes f) = f \circ m \ .$$

4. A <u>morphism $f:A \to A'$ of monoid automata</u> with fixed state object S is a monoid morphism $f:M \to M'$ satisfying
$$d' \circ (f \otimes S) = d \ .$$

5. Monoids and monoid morphisms constitute the category <u>K</u>-<u>Mon</u> of monoids in <u>K</u> and monoid automata together with the morphisms in 4. define the category <u>K</u>-<u>Mon</u>-<u>Aut</u> of monoid automata in <u>K</u> .

<u>Examples</u>: 1. Given an arbitrary <u>K</u>-object I the <u>free monoid</u> $(I^*, m_I, 1_I)$ is defined by the coproduct
$$I^* = \coprod_{n \in \mathbb{N}_o} I^n \qquad \text{with} \quad I^o = U \ , \qquad \mathbb{N}_o = \mathbb{N} \cup \{o\}$$

with coproduct injections $u_n : I^n \to I^*$. The morphism $1_I : U \to I^*$ is exactly u_o and the multiplication $m_I : I^* \otimes I^* \to I^*$ is a generalization of the concatenation of strings which can be defined recursively using the injections u_n.
In the categories $(\underline{Set}, \times)$, $(\underline{PD}, \times)$ and $(\underline{Top}, \times)$ I* coincides with the well-known free monoids whereas we get the free tensor algebra in the case $(\underline{Mod}_R, \otimes)$ (cf. [59]) .

2. Given a Medvedev-automaton $A = (I, S, d)$ with $d: I \otimes S \to S$ we get an extended state transition function, which now is a right action $d^* : I^* \otimes S \to S$, leading to a monoid automaton $A^* = (I^*, S, d^*)$.

3. Using the universal property of the evaluation ev_S (cf. (4.3.1)) we get unique <u>K</u>-morphisms $m_S : \langle S,S \rangle \otimes \langle S,S \rangle \to$ $\to \langle S,S \rangle$ and $1_S : U \to \langle S,S \rangle$ such that the diagrams in (5.6.1) are commutative and $(\langle S,S \rangle, m_S, 1_S)$ becomes a monoid in $(\underline{K}, \otimes)$ called <u>endomorphism monoid</u> of S .

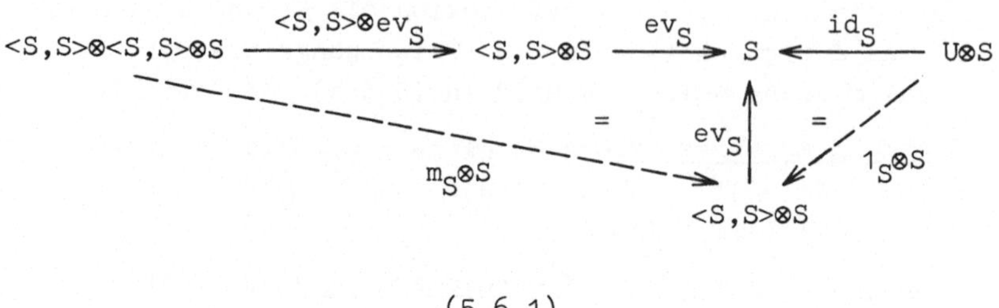

$$(5.6.1)$$

4. Commutativity of (5.6.1) implies that $(<S,S>,S,ev_S)$ is a monoid automaton in $(\underline{K},\otimes)$.

<u>Construction</u> (T r a n s i t i o n   M o n o i d) :
Given a monoid automaton $A = (M,S,d)$ we get a unique $\underline{K}$-morphism $t(A):M \to <S,S>$, called <u>transition morphism</u> of A , such that diagram (5.6.2) is commutative:

$$(5.6.2)$$

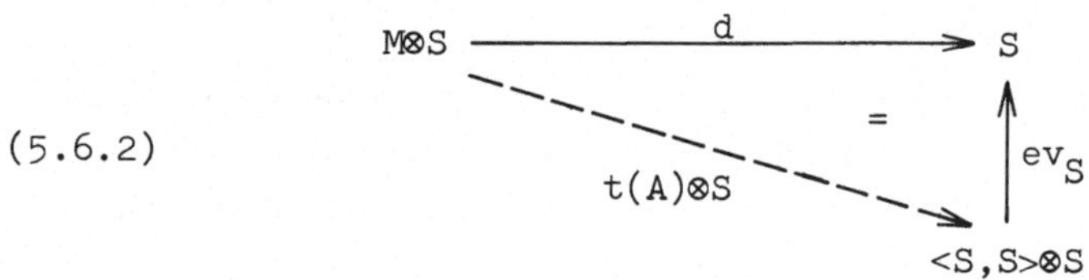

Now the <u>transition monoid</u> $T(A)$ of A is the image of $t(A)$ which analogously to (4.6.2) is given by the canonical $\mathfrak{E}$-$\mathfrak{M}$-factorization (5.6.3) of $t(A)$:

$$(5.6.3)$$

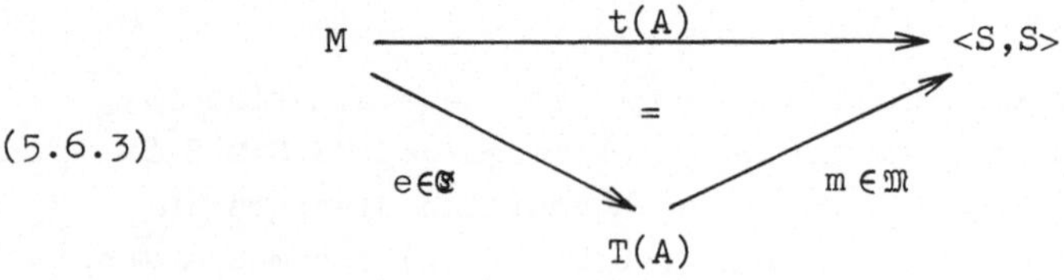

<u>Remarks</u>: The construction of the transition morphism and the
transition monoid is very similar to that of the machine
morphism M(A) in 4.3 and the input-output behavior E(A)
in 4.6 respectively. Moreover it can be shown, similar to
4.5 and 4.8, that t(A) becomes a monoid morphism and T(A)
a submonoid of <S,S>.
Regarding the transition monoid T(A) as the behavior of a
monoid automaton our next problem is to minimize a monoid
automaton with respect to the transition monoid.

<u>5.7 Minimization of Monoid Automata</u>: In order to apply the
general theory of systematics in chapter 3 to monoid automata we first have to extend the construction of the transition monoid T(A) of A to a behavior functor

$$T:\underline{K}\text{-}\underline{Mon}\text{-}\underline{Aut} \longrightarrow \underline{B}$$

where $\underline{B}$ is the category of all submonoids of <S,S> such
that the inclusion belongs to the class $\mathfrak{M}$. This can be
done in exactly the same way as in 5.2 for the input-output
behavior using the fact that in analogy to 4.5 $\underline{K}\text{-}\underline{Mon}\text{-}\underline{Aut}$
is isomorphic to the comma category $(\underline{K}\text{-}\underline{Mon} \downarrow <S,S>)$ which
has as objects monoid morphisms $t:M \rightarrow <S,S>$ with arbitrary
monoids M in $\underline{K}$ and morphisms in $(\underline{K}\text{-}\underline{Mon} \downarrow <S,S>)$ are
monoid morphisms $f:M \rightarrow M'$ satisfying $t' \circ f = t$.
Thus we get a systematic $\underline{K}\text{-}\underline{Mon}\text{-}\underline{Aut} = (\underline{K}\text{-}\underline{Mon}\text{-}\underline{Aut},\underline{B},T)$ of
monoid automata in the sense of 3.1. Using the same con-
structions for $\underline{K}\text{-}\underline{Mon}\text{-}\underline{Aut}$ as we have done for the systematic $\underline{K}\text{-}\underline{Aut}$ in 5.3, we obtain the following strong result:

<u>Theorem</u> (M i n i m i z a t i o n   o f   t h e
T r a n s i t i o n   M o n o i d) : The systematic
$\underline{K}\text{-}\underline{Mon}\text{-}\underline{Aut}$ of monoid automata satisfies the Minimal Realization Principle.
In more detail we have by 3.6 : The class of all "observable"
monoid automata A , meaning that the transition morphism
t(A) belongs to the class $\mathfrak{M}$, defines a minimal, reduced
and realizing subsystematic of $\underline{K}\text{-}\underline{Mon}\text{-}\underline{Aut}$. Hence there is

for each monoid automaton $A = (M,S,d)$ an "equivalent" minimal automaton $A' = (M',S,d')$ such that $T(A) = T(A') = M'$
and d' is the restriction of $ev_S:<S,S>\otimes S \to S$ to $M'\otimes S$.
Moreover $e \in \mathcal{C}$ in (5.6.3) is a "reduction" from A to A' ,
i.e. a morphism of monoid automata which belongs to the
class $\mathcal{C}$.

Thus the systematic <u>K-Mon-Aut</u> is another example which can
be applied to the general theory of chapter 3 but it is left
to the reader to interprete all the other results of 3.4 and
3.6.

<u>Remark</u> (I n p u t - O u t p u t - B e h a v i o r   o f
M o n o i d   A u t o m a t a) : Considering monoid automata with output we can once again ask for the input-output-
behavior which has before been studied for automata of
Mealy-type. In the case of monoid automata we will simply
add an output morphism $l:S \to O$ which corresponds to the
Moore-type (cf. 1.14). Then it is easy to check that all
constructions and results of the chapters 4 and 5 can be extended to the case of monoid automata of Moore-type replacing
only automata by monoid automata with output $l:S \to O$,
Medvedev-automata by monoid automata and $<I^+,O>$ by $<M,O>$
respectively. Of course, we do not fix the state object S
in this case but the input monoid M and output object O .
Moreover we assume that the state transition morphism
$d:S\otimes M \to S$ is again a right action.

Finally let us remark that we have meanwhile generalized
theorems 2.2, 2.4 and the constructions in 2.8 and specialized theorems 3.4 and 3.6 to automata in closed categories.
Moreover we have solved all the problems stated in 2.6.
The characterization of finite minimal realizations is given
in the corollary of 4.8 combined with theorem 5.3.

Thus we have shown that a great number of problems and results stated for deterministic automata can be extended to
automata in closed categories and thus to all our examples

given in 4.9. Automata of nondeterministic type will be
studied in the following chapters, and in chapter 9 we will
extend the theory for both types to automata with fixed
initial state.

6. Behavior of Automata in Pseudoclosed Categories: The Nondeterministic Case

In the chapters 4 and 5 we have studied automata in closed
categories, basing on the example of deterministic automata.
Now we are going to generalize nondeterministic automata to
automata in pseudoclosed categories. The construction of the
extended output morphism $1^+:S \otimes I^+ \to O$ is exactly the same
as before but unfortunately, the category $(\underline{ND}, \times)$ for example
is not closed such that we do not get the machine morphism
$M(A)$ as an adjoint morphism of 1^+ in $\underline{ND}$ (cf. (4.3.3)).
On the other hand we know that the nondeterministic function
$1^+:S \times I^+ \to O$ can be regarded as a deterministic function
from $S \times I^+$ to $\wp'O$ which is the powerset of O without
the empty set. Hence we get a deterministic machine function
$M(A):S \to <I^+, \wp'O>$ (cf. 2.7). In fact the main idea of pseudoclosed categories $(\underline{K}, \otimes)$ is to represent "nondeterministic"
morphisms $f:A \to B$ in $\underline{K}$ as "deterministic" morphisms
$f':A \to PB$ in $\underline{K}'$ where $(\underline{K}', \otimes)$ is a suitable closed monoidal subcategory of $(\underline{K}, \otimes)$ and PB a suitable object in
$\underline{K}'$. In other words we assume that $(\underline{K}, \otimes)$ has a coreflexive
closed monoidal subcategory $(\underline{K}', \otimes)$. In this case $(\underline{K}, \otimes)$
will be called "pseudoclosed". Indeed the categories $(\underline{ND}, \times)$,
$(\underline{Rel}, \times)$ and $(\underline{Stoch}, \times)$ are pseudoclosed relative $(\underline{Set}, \times)$
and $(\underline{RelTop}, \otimes)$ relative $(\underline{Top}, \otimes)$ such that our theory will
be applicable to nondeterministic, relational, stochastic
and relational topological automata. A list of other examples
will be given in 6.8.

As in chapter 4 we will construct the extended output morphism, machine morphism and behavior for automata in pseudoclosed categories, together with corresponding interpretations for our examples. The main results will be a characterization of the machine morphism and of the behavior in 6.6

and 6.7 respectively. Reduction, minimization and realization will be studied in chapter 7.

6.1 <u>General Assumptions</u>: In this chapter $(\underline{K},\otimes)$ and $(\underline{K}',\otimes)$ will be monoidal categories such that $(\underline{K}',\otimes)$ is a monoidal subcategory of $(\underline{K},\otimes)$ having the same class of objects. Especially we assume that the tensor product in $\underline{K}'$ is the restriction of the tensorproduct in $\underline{K}$ such that we can take the same symbol $\otimes$. Moreover we assume that
(i) $(\underline{K}',\otimes)$ is closed, has countable coproducts and an $\mathfrak{E}$-$\mathfrak{M}$-factorization with canonical representatives in $\mathfrak{M}$ (cf. 4.1)
(ii) $(\underline{K},\otimes)$ is pseudoclosed relative $(\underline{K}',\otimes)$ with right adjoint $P:\underline{K} \rightarrow \underline{K}'$ of the inclusion functor $J:\underline{K}' \rightarrow \underline{K}$. The definition of pseudoclosed categories will be given below in 6.2:

6.2 <u>Pseudoclosed Categories</u>: Generalizing the bijective correspondence between nondeterministic functions $f:A \rightarrow B$ and deterministic functions $f':A \rightarrow P'B$ in the closed category $(\underline{Set},x)$ we get:

<u>Definition</u>: A monoidal category $(\underline{K},\otimes)$ is called <u>pseudoclosed</u> relative a closed monoidal subcategory $(\underline{K}',\otimes)$ if the inclusion functor $J:\underline{K}' \rightarrow \underline{K}$ has a right adjoint functor $P:\underline{K} \rightarrow \underline{K}'$. In more detail, this condition asserts that for each object 0 in $\underline{K}$ there is a morphism $v:P0 \rightarrow 0$ in $\underline{K}$, called counit for 0, such that for each $\underline{K}$-morphism $f:A \rightarrow 0$ there is a unique $\underline{K}'$-morphism $f':A \rightarrow P0$ in (6.2.1) satisfying $v \circ f' = f$.

(6.2.1)

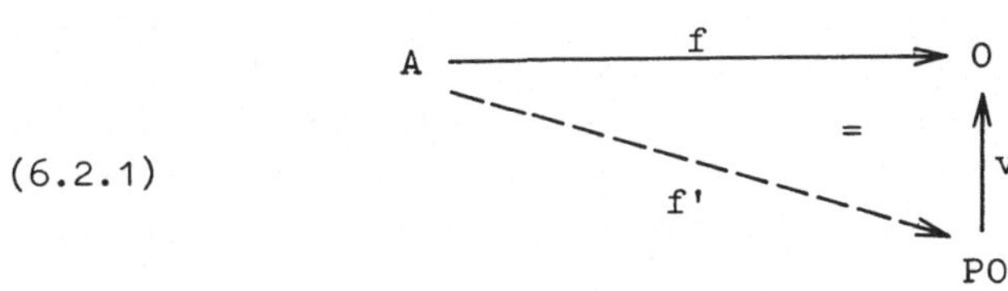

In other words (6.2.1) means that $\underline{K}'$ is a _coreflexive sub-category_ of $\underline{K}$, and $v:PO \to O$, or more precisely $v_O:PO \to O$, is the counit of the adjunction $J \dashv P$ for the object O (cf. 12.7).

Combining this property with the assumption that $(\underline{K}',\otimes)$ is closed we get the following important lemma which is a weaker analogue to lemma 4.4 in the closed case:

<u>Lemma</u>: Let $(\underline{K},\otimes)$ be pseudoclosed relative $(\underline{K}',\otimes)$ then we have for each object K in $\underline{K}'$ a right adjoint functor $<K,P->:\underline{K} \to \underline{K}'$ of the functor $J \circ (-\otimes K):\underline{K}' \to \underline{K}$.
In more detail, the composition $v \circ ev:<K,PO>\otimes K \to O$ of the evaluation ev and the morphism v as given in (6.2.1) has the following property:
For each object S in $\underline{K}'$ and each $\underline{K}$-morphism $f:S\otimes K \to O$ there is a unique $\underline{K}'$-morphism $f':S \to <K,PO>$ in (6.2.2) satisfying $(v \circ ev) \circ (f' \otimes K) = f$.

(6.2.2)

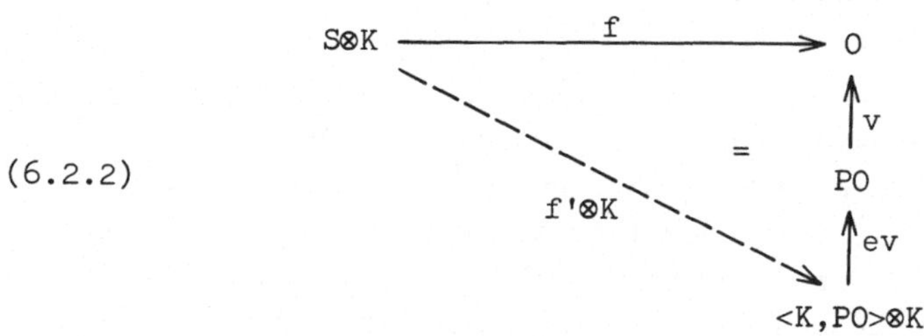

<u>Proof</u>: First there is a unique $\underline{K}'$-morphism $f_1:S\otimes K \to PO$ such that $v \circ f_1 = f$ using the coreflexivity of $\underline{K}'$ and then we have a unique $\underline{K}'$-morphism $f':S \to <K,PO>$ satisfying $ev \circ (f' \otimes K) = f_1$ because $(\underline{K}',\otimes)$ is closed. Thus f' satisfies $v \circ ev \circ (f' \otimes K) = v \circ f_1 = f$ and is unique with respect to this property which follows directly from the uniqueness properties of f_1 and f' . ∎

In categorical terms we simply have a composition of adjoint functors leading again to adjoint functors (cf. 12.7).

<u>Examples</u>: 1. The category $(\underline{ND},\times)$ of nondeterministic func-
tions (cf. 1.10,2) is pseudoclosed relative the closed cate-
gory $(\underline{Set},\times)$ (cf. 4.3,1). As explained in the introduction
we take $PO := P'O$ to be the powerset of O without the empty set and $v:PO \to O$ is the nondeterministic function assigning to each non-empty subset O' of O the set of all
elements of O', i.e. O' itself. Now given $f:A \to O$ in $\underline{ND}$
$f':A \to PO$ is defined by $f'(a) = f(a)$ for all $a \in A$ since
$f(a)$ is already a non-empty subset of O by 1.4. Clearly
we have $v \circ f' = f$ and all functions $f'':A \to PO$ satisfying
$v \circ f'' = f$ are equal to f' .

2. Similar to example 1 the category $(\underline{Rel},\times)$ of relations
(cf. 1.13,2) is pseudoclosed relative $(\underline{Set},\times)$ taking PO to
be the powerset of O and $v:PO \to O$ is defined by $O'v\,x$
iff $x \in O'$ for all $O' \in PO$ and $x \in O$. The argument for showing that (6.2.1) is satisfied is nearly the same as in the
first example.

3. Finally the category $(\underline{Stoch},\times)$ of stochastic channels is
a third category which is pseudoclosed relative $(\underline{Set},\times)$.
According to 1.10,3 $(\underline{Set},\times)$ is a monoidal subcategory of
$(\underline{Stoch},\times)$ and by 1.5 each stochastic channel $f:A \to O$ can
be regarded as a function $f':A \to PO$ where PO is the set
of all (discrete) probability distributions p on O .
Defining the stochastic channel $v:PO \to O$ by $v(p)(x) = p(x)$
for all $p \in PO$ and $x \in O$ we have $v \circ f' = f$ using the composition defined in 1.10,3 :

$$(v \circ f')(a)(x) = \sum_{p \in PO} [f'(a)(p)] \cdot [v(p)(x)]$$
$$= 1 \cdot v(f'(a))(x) = f'(a)(x) = f(a)(x)$$

Note that $f'(a)(p) = 1$ only in the case $p = f'(a)$ and zero
otherwise. The same formula proves that f' is the unique
function $f':A \to PO$ satisfying $v \circ f' = f$.

4. The category $(\underline{RelTop},\otimes)$ defined in 1.13,3 is pseudoclosed

relative $(\underline{Top},\otimes)$ taking PO to be the powerset $\mathcal{P}O$ of O with the following "lower finite" topology: Given an open set U in O an element of a subbasis of $\mathcal{P}'O$ contains all subsets V of O such that the intersection $V\cap U$ is not empty. If U ranges over all open sets of O we get a subbasis of $\mathcal{P}'O$. $\mathcal{P}O$ is assumed to have the minimal topology such that the inclusion $\mathcal{P}'O \subseteq \mathcal{P}O$ is an open mapping. Now we are able to show that a relation $f:A \to O$ is lower semi-continuous iff the corresponding function $f':A \to PO$ is continuous. Note that it suffices to test the continuity of f' on the subbasis elements $\mathfrak{B} = \{V/\ V\cap U \neq \phi\}$ where U is an arbitrary open set in O . Thus we have

$$f'^{-1}(\mathfrak{B}) = \{a\in A/\ f'(a)\in\mathfrak{B}\} = \{a\in A/\ f'(a) \cap U \neq \phi\}$$

$$= \{a\in A/\ f(a) \cap U \neq \phi\} = f^{-1}(U)$$

showing that $f^{-1}(U)$ is open iff $f'^{-1}(\mathfrak{B})$ is open in A or f is lower semi-continuous iff f' is continuous. Especially the relation $v:PO \to O$ defined already in example 2 is lower semi-continuous because $v':PO \to PO$ is the identity. Hence by example 2 we are sure that (6.2.1) is satisfied and thus $(\underline{RelTop},\otimes)$ is pseudoclosed relative $(\underline{Top},\otimes)$ which was already shown to be closed in 4.3 example 4 .

Remark: Corresponding to our situation of pseudoclosed categories morphisms of the form $f:A \to PO$ are called fuzzy morphisms in [6], written $f:A \to O$, taking our bijective correspondence in (6.2.1) between $\underline{K}$-morphisms $f:A \to O$ and $\underline{K}'$-morphisms $f':A \to PO$ as the basic definition. Adding some axioms for the composition of fuzzy morphisms they define a fuzzy theory which is shown to be in bijective correspondence to "monads" in $\underline{K}'$ (cf. [59]). On the other hand given a category $\underline{K}$ such that $\underline{K}'$ is a coreflexive subcategory, which we have in the case that $(\underline{K},\otimes)$ is pseudoclosed relative $(\underline{K}',\otimes)$, the composition $P\circ J:\underline{K}' \to \underline{K}'$ of the inclusion $J:\underline{K}' \to \underline{K}$ and the right adjoint $P:\underline{K} \to \underline{K}'$ defines a monad in $\underline{K}'$ and thus corresponds to a fuzzy the-

ory in the sense of [6]. Due to the remark in 4.4, showing
that automata in closed categories correspond to adjoint machines in the sense of [5], the approach in [6] of nondeterministic machines based on fuzzy morphisms corresponds to our
theory of automata in pseudoclosed categories. An approach
similar to that in [6] has been given independently in [19]
but up to now both approaches are only sketched.

<u>6.3 Automata in Pseudoclosed Categories</u>: Given an automaton
$A = (S,d,l)$ in the pseudoclosed category $(\underline{K},\otimes)$ we are going
to construct the extended output morphism, the machine mor-
phism and the behavior of A .

1. (E x t e n d e d   O u t p u t   M o r p h i s m) :
The construction of $1^+:S\otimes I^+ \to O$ is the same as given in
4.2. Actually in that section we did not use that $(\underline{K},\otimes)$ is
closed, but only that $(\underline{K},\otimes)$ has countable coproducts which
are compatible with $\otimes$ (cf. assumption in 4.2). Since we
have this assumption for $(\underline{K}',\otimes)$ and $\underline{K}'$ has the same class
of objects as $\underline{K}$ it remains to show:

<u>Lemma</u>: Each coproduct K with injections $u_n:K_n \to K$ $(n\in\mathbb{N})$
in $\underline{K}'$ is already a coproduct in $\underline{K}$.

As a corollary we get that $S\otimes K$ with injections
$S\otimes u_n:S\otimes K_n \to S\otimes K$, which is a coproduct in $\underline{K}'$ by 4.2, is also
a coproduct in $\underline{K}$. The proof of the lemma follows from the
fact that the inclusion $J:\underline{K}' \to \underline{K}$ is a left adjoint functor
and hence preserves colimits and especially coproducts
(cf. 12.9). Nevertheless we give a direct proof of the lemma
which is very similar to that in 6.2 :

<u>Proof of the Lemma</u>: Given a family of $\underline{K}$-morphisms $f_n:K_n \to O$
$(n\in\mathbb{N})$ there are unique $\underline{K}'$-morphisms $f_n':K_n \to PO$ satisfying
$v\circ f_n' = f_n$ because $\underline{K}'$ is coreflexive in $\underline{K}$. Using the coproduct properties of K in $\underline{K}'$ there is a unique $f':K\to PO$
in $\underline{K}'$ such that $f'\circ u_n = f_n'$ for all $n\in\mathbb{N}$. Now it is easy
to see that $f := v\circ f':K \to O$ is the unique $\underline{K}$-morphism satisfying $f\circ u_n = f_n$ for all $n\in\mathbb{N}$.   ∎

Now let us recall from the construction in 4.2 that the output morphism $1:S\otimes I \to O$ leads to a family $1_n:S\otimes I^n \to O$ $(n\in\mathbb{N})$ of $\underline{K}$-morphisms defined by

$$(6.3.1) \qquad 1_1 = 1 \quad , \quad 1_{n+1} = 1_n \circ (d\otimes I^n) \quad (n\in\mathbb{N}) \ .$$

By assumption, $S\otimes I^+$ is a coproduct in $\underline{K}'$ with injections $S\otimes i_n:S\otimes I^n \to S\otimes I^+$ (cf. 4.2) but this is also a coproduct in $\underline{K}$ by the above lemma. Thus the <u>extended output</u> is the unique $\underline{K}$-morphism $1^+:S\otimes I^+ \to O$ in the following commutative diagram

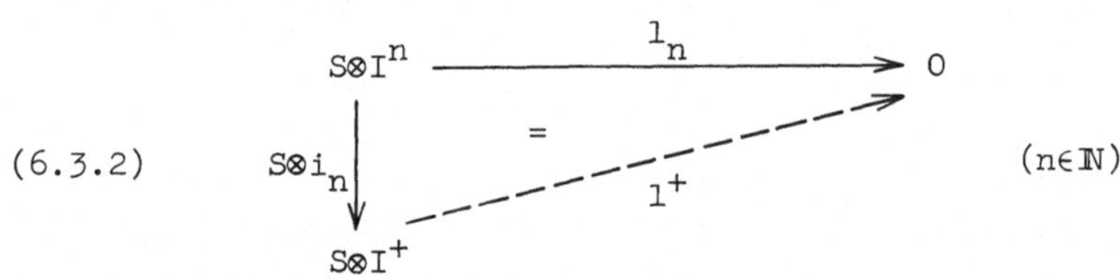

$$(6.3.2) \hspace{6em} (n\in\mathbb{N})$$

2. (M a c h i n e   M o r p h i s m) : Using (6.2.2) the extended output $1^+:S\otimes I^+ \to O$, which is a $\underline{K}$-morphism, leads to a unique $\underline{K}'$-morphism $M(A):S \to \langle I^+,PO\rangle$ in (6.3.3), called <u>machine morphism</u> of A, such that $(v\circ ev)\circ(M(A)\otimes I^+) = 1^+$.

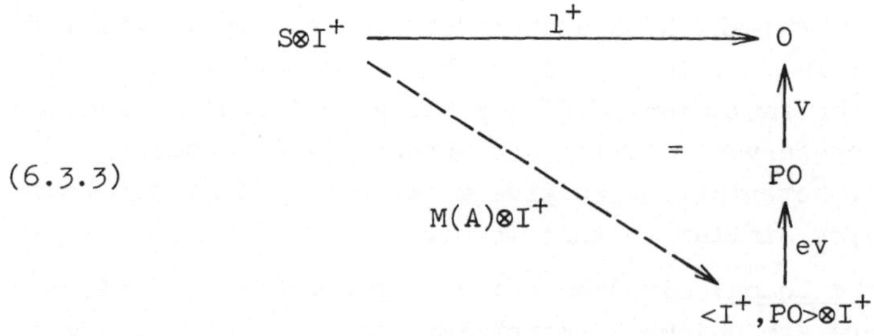

$$(6.3.3)$$

This construction of $M(A)$ corresponds to (4.3.3) for automata in closed categories. Note that in our case the $\underline{K}$-morphism 1^+ leads to a morphism $M(A)$ in the closed category $(\underline{K}',\otimes)$.

3. (B e h a v i o r) : Since the machine morphism $M(A)$ of A is a $\underline{K}'$-morphism we get the behavior of A using the canonical $\mathfrak{E}$-$\mathfrak{M}$-factorization $m(A) \bullet e(A) = M(A)$ of $M(A)$ in (6.3.4) which exists by assumption 6.1,(i).

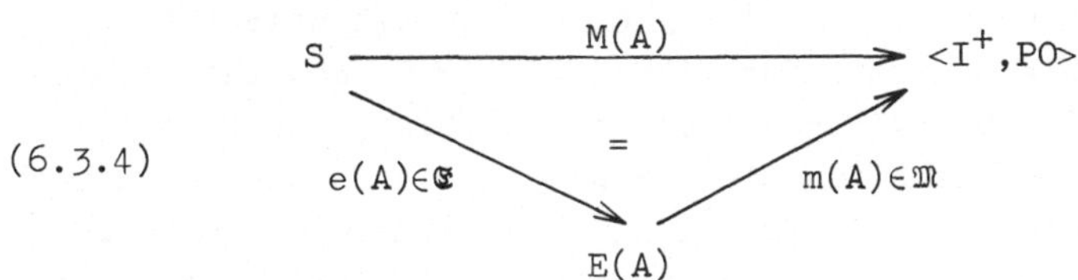

The image $m(A):E(A) \to \, <I^+,PO>$ of $M(A)$, or for short $E(A)$, is called <u>behavior</u> of A .
The only difference to the construction in 4.6 for automata in closed categories is the fact that O is now replaced by PO . But we will see in 6.4 that the important characterization theorem 4.5 of $\underline{K}$-$\underline{Aut}$ using the machine morphisms does not remain true in the pseudoclosed case.

<u>Interpretation</u>: According to our examples of pseudoclosed categories in 6.2 the above constructions can be interpreted as follows:

1. In the case of nondeterministic automata $(\underline{K}, \otimes)$ is the category $(\underline{ND}, \times)$ which is pseudoclosed relative $(\underline{Set}, \times)$ (cf. 6.2 example 1). By (6.3.2) and (6.3.1) and the definition of the composition in $\underline{ND}$ (cf. 1.10,2) we have for all $s \in S$, $x \in I$, $w \in I^n$

$$1^+(s,xw) = 1_{n+1}(s,xw) = 1_n \circ (d \times I^n)((s,x),w)$$

$$= \bigcup_{s' \in d(s,x)} 1_n(s',w) \, .$$

Thus the last output $1^+(s,xw)$ induced by the input string xw starting with the state $s \in S$ is the union of all last output sets $1_n(s',w)$ ranging over all possible next states $s' \in d(s,x)$. In other words $y \in O$ is one of the last output

symbols induced by the string $xw = x_1 \ldots x_{n+1}$ starting with
state s iff there is a sequence $s_1, \ldots, s_n$ of states in
S satisfying $s_1 \in d(s,x_1)$, $s_2 \in d(s_1,x_2)$, $\ldots$, $s_n \in d(s_{n-1},x_n)$
and $y \in l(s_n,x_{n+1})$.
Since PO is equal to $\wp'O$ in our example $\langle I^+,PO \rangle$ is the
set of all functions $f':I^+ \to \wp'O$ or equivalently the set
of all nondeterministic functions $f:I^+ \to O$ and the machine
morphism $M(A):S \to \langle I^+,PO \rangle$ defined by (6.3.3) assigns to
each state $s \in S$ the function $M(A)(s):I^+ \to PO$ corresponding
to the nondeterministic function $1^+(s,-):I^+ \to O$ which is
the input-output behavior of the state s . This is easy to
verify because we have for each $s \in S$ and $w \in I^+$ by (6.3.3):

$$1^+(s,w) = (v \circ ev) \circ (M(A) \times I^+)(s,w) = (v \circ ev)(M(A)(s),w)$$

$$= v(M(A)(s)(w)) = M(A)(s)(w)$$

(cf. (2.1.1) and example 6.2,1 for the definition of ev and
v respectively).
Finally the behavior $E(A)$ of A defined as the image of
$M(A)$ is the set

$$E(A) = \{M(A)(s):I^+ \to \wp'O \ / \ s \in S\}$$

of all input-output behaviors of the states $s \in S$.

2. The case of relational automata can be treated similarly
to example 1 . We only have to replace $\underline{ND}$ by $\underline{Rel}$, i.e.
nondeterministic functions by relations which can be regarded
as partially defined, nondeterministic functions (cf. 1.13,2).
Thus replacing $\wp'O$ by the powerset PO we have the same
interpretation of 1^+ , $M(A)$ and $E(A)$ as in the nondeterministic case where now the sets $1^+(s,w) = M(A)(s)(w)$ may
be empty and $1^+(s,-):I^+ \to O$ may be the empty relation.

<u>Remark</u>: There is another possibility for the construction of
the machine morphism in the case of relational automata because the relation $1^+:S \times I^+ \to O$ defines a unique relation
$M:S \to I^+ \times O$ given by $(s,(w,y)) \in M$ iff $((s,w),y) \in 1^+$. Now
the relation M can be regarded as a function $M':S \to \wp(I^+ \times O)$
which coincides with the machine morphism $M(A):S \to \langle I^+,PO \rangle$

because $\mathcal{P}(I^+\times O)$ is the set of all relations from I^+ to O.
The construction of the relation $M:S \to I^+\times O$ is exactly the
same as in (4.3.3) for the case of automata in closed categories because the category $(\underline{Rel},\times)$ actually is a closed
category with $K\times-$ right adjoint to $-\times K$ (cf.4.3). But the
category $(\underline{Rel},\times)$ has no $\mathfrak{C}$-$\mathfrak{M}$-factorization such that the
theory of automata in closed categories is not applicable to
relational automata. That is not only a technical difficulty
because several of the results given in chapter 5, e.g.
uniqueness of minimization, are well known to be false for
nondeterministic automata (cf. 2.7).

3. Regarding stochastic automata, PO is the set of all discrete probability distributions on O (cf. 6.2 example 3)
such that $<I^+,PO>$ can be regarded as the set of all stochastic channels $f:I^+ \to O$. According to (6.3.3) the machine
morphism $M(A):S \to <I^+,PO>$ is a function assigning to each
state $s\in S$ the function $M(A)(s):I^+ \to PO$ corresponding to
the stochastic channel $1^+(s,-):I^+ \to O$, i.e. the input-output channel of the state s. Finally the behavior $E(A)$ is
the set of all these input-output channels ranging over all
$s\in S$.

4. The interpretation for relational topological automata is
the same as that given for relational automata, but with the
additional properties that $1^+:S\otimes I^+ \to O$ becomes a lower
semi-continuous relation, $M(A):S \to <I^+,PO>$ a continuous
function and the behavior $E(A)$ carries subspace topology
of $<I^+,PO>$ in the case that $\mathfrak{M}$ is the class of all extremal
monomorphisms in $\underline{Top}$ (cf. 4.6 example 4). The space $<I^+,PO>$
of all continuous functions $f':I^+ \to PO$, which can be regarded as lower semi-continuous relations $f:I^+ \to O$, is endowed
with the topology of pointwise convergence (cf. 4.3 example 4).

<u>6.4 Example and Motivation</u> (U n i o n) : As promised, we
want to give an example of a nondeterministic automaton in

order to calculate 1^+, $M(A)$ and $E(A)$ as defined in 6.3. We will see that the machine morphism $M(A):S \to \langle I^+,\wp'O\rangle$ is not compatible with $d:S\times I \to S$ and $L:\langle I^+,\wp'O\rangle\times I \to \langle I^+,\wp'O\rangle$ in general, as it was in the deterministic case. More precisely we do not have $M(A)\circ d = L\circ(M(A)\times I)$ (cf. corollary of 4.5) unless the automaton $A = (S,d,l)$ has the property that for each $s\in S$ and $x\in I$ all the next states $s_1',s_2'\in d(s,x)$ have the same input-output morphism $M(A)(s_1') = M(A)(s_2'):I^+ \to \wp'O$.

Given $I = \{x\}$ and $O = \{y_1,y_2\}$ we consider the nondeterministic automaton $A_5 = (S_5,d_5,l_5)$ with $S_5 = \{1,2,3\}$, $d_5(1,x) = \{2\}$, $d_5(2,x) = \{1\}$, $d_5(3,x) = \{1,2\}$ and $l_5(1,x) = \{y_1\}$, $l_5(2,x) = \{y_2\}$, $l_5(3,x) = \{y_1,y_2\}$ given by the following diagram

(6.4.1)

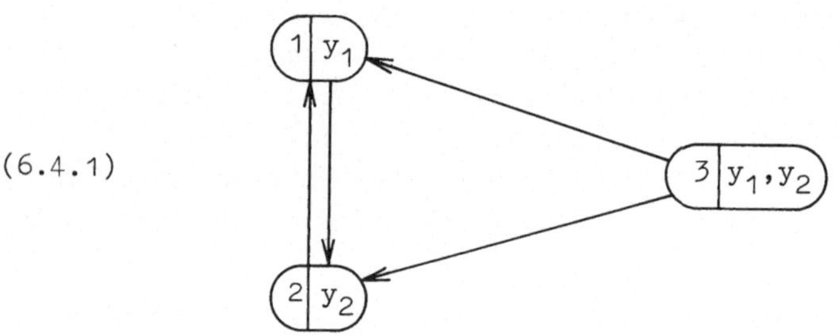

Calculation of $l_5^+:S_5\times I^+ \to O$ and $M(A_5):S_5 \to \langle I^+,\wp'O\rangle$ yields:
$$l_5^+(1,x^n) = M(A_5)(1)(x^n) = \left\{ \begin{array}{ll} \{y_1\} & \text{if } n \text{ is odd} \\ \{y_2\} & \text{if } n \text{ is even} \end{array} \right.$$

$$l_5^+(2,x^n) = M(A_5)(2)(x^n) = \left\{ \begin{array}{ll} \{y_2\} & \text{if } n \text{ is odd} \\ \{y_1\} & \text{if } n \text{ is even} \end{array} \right.$$

$$l_5^+(3,x^n) = M(A_5)(3)(x^n) = \{y_1,y_2\} \text{ for all } n\geq 1 \quad .$$

The behavior $E(A_5) = \{f_1,f_2,f_3\}$ of A_5 consists of the functions $f_i = M(A_5)(i):I^+ \to \wp'O$ $(i = 1,2,3)$.
Now we consider the condition $M(A_5)\circ d_5 = L\circ(M(A_5)\times I)$ for state 3 and input x:

$$M(A_5)\circ d_5(3,x) = M(A_5)(\{1,2\}) = \{f_1,f_2\}$$

$$L \circ (M(A_5) \times I)(3,x) = L(M(A_5)(3),x) = M(A_5)(3) \circ L_x = f_3 \circ L_x = f_3$$

Hence the above condition is not satisfied, but taking the union $f_1 \cup f_2$ defined by $(f_1 \cup f_2)(x^n) = f_1(x^n) \cup f_2(x^n)$ we get $f_1 \cup f_2 = f_3$. This motivates to modify the above condition. Note that the union can be regarded as a function

(6.4.2) $\qquad\qquad u:\wp'\langle I^+,\wp'0\rangle \to \langle I^+,\wp'0\rangle$

defined by $u(F) = \bigcup_{f \in F} f$ for each $F \in \wp'\langle I^+,\wp'0\rangle$ which is a subset of all functions $f:I^+ \to \wp'0$.

The modified condition turns out to be (cf. 6.6.1) :

(6.4.3) $\qquad\qquad u \circ \wp'M(A) \circ d' = L \circ (M(A) \times I)$

where $d':S \times I \to \wp'S$ is the function associated with $d:S \times I \to S$ and $\wp'M(A):\wp'S \to \wp'\langle I^+,\wp'0\rangle$ is the extension of $M(A)$ to subsets.

The construction of the union u and the condition (6.4.3) can be generalized to automata in pseudoclosed categories, this will be done in 6.5 and 6.6 .

<u>6.5 Definition</u> (U n i o n) : Given a pseudoclosed category $(\underline{K},\otimes)$ relative $(\underline{K}',\otimes)$ with right adjoint $P:\underline{K} \to \underline{K}'$ and the counit morphisms $v:P0 \to 0$ and $v_1:P\langle I^+,P0\rangle \to \langle I^+,P0\rangle$ for the objects 0 and $\langle I^+,P0\rangle$ (cf. 6.2) there is a unique $\underline{K}'$-morphism $u:P\langle I^+,P0\rangle \to \langle I^+,P0\rangle$, called <u>union</u>, such that diagram (6.5.1) commutes.

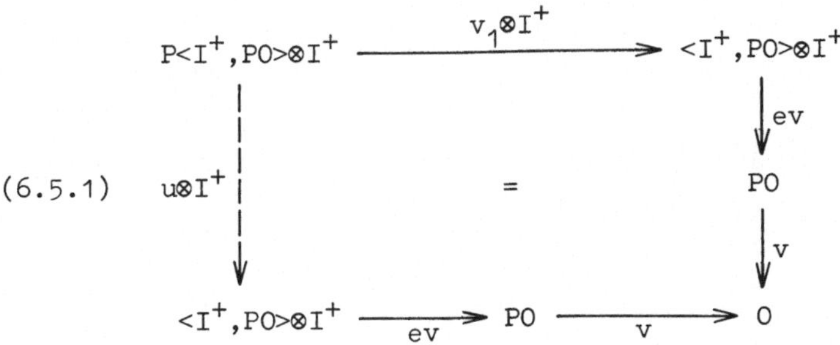

Existence and uniqueness of u follow from (6.2.2) taking $f = v \circ ev \circ (v_1 \otimes I)$. Note that v_1 and u have the same domain and codomain but v_1 is a $\underline{K}$-morphism and u is a morphism in the closed category $\underline{K}'$.

$\underline{\text{Interpretation}}$: The morphism $u: P\langle I^+, PO \rangle \to \langle I^+, PO \rangle$ corresponds to the union of nondeterministic functions in the case $(\underline{ND}, \times)$. Similarly u is the union of relations in the examples $(\underline{Rel}, \times)$ and $(\underline{RelTop}, \otimes)$ and a weighted sum of probability distributions in the case $(\underline{Stoch}, \times)$.

1. In order to show that u is the union in the category $(\underline{ND}, \times)$ we apply (6.5.1) to a subset $F \in P\langle I^+, PO \rangle$ of functions $f: I^+ \to PO$ and $w \in I^+$ using the fact that $u: P\langle I^+, PO \rangle \to \langle I^+, PO \rangle$ is a deterministic function:

$$[(v \circ ev) \circ (u \times I^+)](F,w) = (v \circ ev)(u(F),w) = \underline{u(F)(w)}$$
$$= [(v \circ ev) \circ (v_1 \times I^+)](F,w) = \bigcup_{f \in v_1(F)} (v \circ ev)(f,w) = \underline{\bigcup_{f \in F} f(w)}$$

Thus we have $u(F)(w) = \bigcup\limits_{f \in F} f(w)$ showing that u is the union defined in 6.4 .

2. In the category $(\underline{Stoch}, \times)$, $F \in P\langle I^+, PO \rangle$ is a probability distribution on the set of all functions $f': I^+ \to PO$ which are stochastic channels $f: I^+ \to O$ (cf. 6.2 example 3). Thus $u(F) \in \langle I^+, PO \rangle$ is a stochastic channel $u(F): I^+ \to O$ which turns out to be the following weighted sum of all stochastic channels $f: I^+ \to O$:

$$u(F) = \sum_{f \in \langle I^+, PO \rangle} F(f) \cdot f$$

because (6.5.1) applied to $(F,w) \in P\langle I^+, PO \rangle \times I^+$ and $y \in O$ yields:

$$[(v \circ ev) \circ (u \times I^+)](F,w)(y) = [(v \circ ev)(u(F),w)](y) = \underline{u(F)(w)(y)}$$
$$= [(v \circ ev) \circ (v_1 \times I^+)](F,w)(y)$$
$$= \sum_{(f,w') \in \langle I^+, PO \rangle \times I^+} (v_1 \times I^+)(F,w)(f,w') \cdot (v \circ ev)(f,w')(y)$$
$$= \sum_{f \in \langle I^+, PO \rangle} v_1(F)(f) \cdot (v \circ ev)(f,w)(y) = \underline{\sum_{f \in \langle I^+, PO \rangle} F(f) \cdot f(w)(y)} \ .$$

<u>6.6 Theorem</u> (C h a r a c t e r i z a t i o n   o f   M a -
c h i n e   m o r p h i s m s) : A $\underline{K}$'-morphism $g:S \to \langle I^+,PO\rangle$
is the machine morphism of an automaton A , i.e. $g = M(A)$
(cf. (6.3.3)), iff there is a $\underline{K}$'-morphism $d':S\otimes I \to PS$ such
that the following diagram (6.6.1) is commutative in $\underline{K}$' :

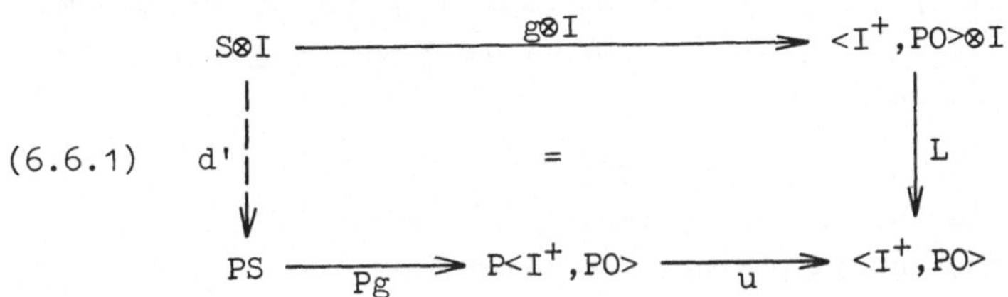

$$(6.6.1)$$

where L is the left shift morphism of PO (cf. (4.4.1)),
       u is the union defined in (6.5.1),
       Pg according to (6.2.1) is the unique $\underline{K}$'-morphism
satisfying

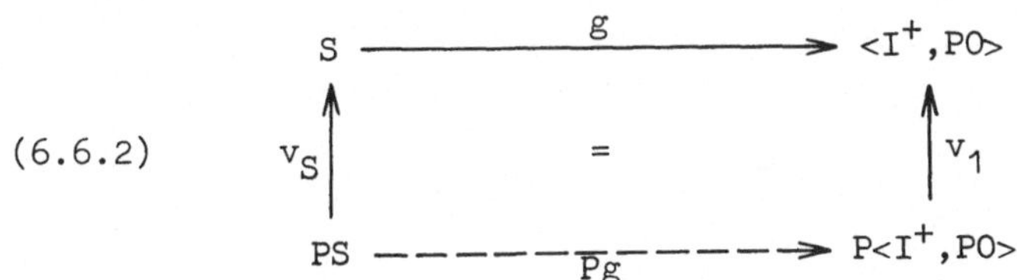

$$(6.6.2)$$

with v_1 and v_S being the counits of the objects $\langle I^+,PO\rangle$
and S (cf. (6.2.1)). In other words Pg is the applica-
tion of the functor P to g.

<u>Remark</u>: Taking $d := v_S \circ d':S\otimes I \to S$ the condition $g \circ d =$
$= L \circ (g\otimes I)$ is not in general satisfied by machine morphisms
g as it is in the case of automata in closed categories
(cf. corollary of 4.5). But we have the weaker condition

$$(6.6.3) \qquad (v \circ ev) \circ (g \otimes I^+) \circ (d \otimes I^+) = (v \circ ev) \circ (L \otimes I^+) \circ (g \otimes I \otimes I^+)$$

which is equivalent to (6.6.1). This can be directly seen
using the diagram (6.6.5) below.
Similar to theorem 4.8 we obtain a behavior characterization
in 6.7 as a corollary of this theorem.

<u>Proof</u>: Let $g = M(A)$ be the machine morphism of the automa-
ton $A = (S,d,l)$. By (6.2.1) there is a unique $\underline{K}'$-morphism
$d':S \otimes I \to PS$ satisfying

$$(6.6.4) \qquad\qquad\qquad v_S \circ d' = d .$$

According to (6.2.2) the commutativity of (6.6.1) is equivalent to the condition that the diagram (6.6.1) tensored
with I^+ is equalized by the counit $v \circ ev$. But this condition is satisfied because the diagram (6.6.5) below is commutative which follows from the corresponding references
($\otimes$ means that the respective diagram is commutative because
the tensor product is a bifunctor, cf. 12.5).

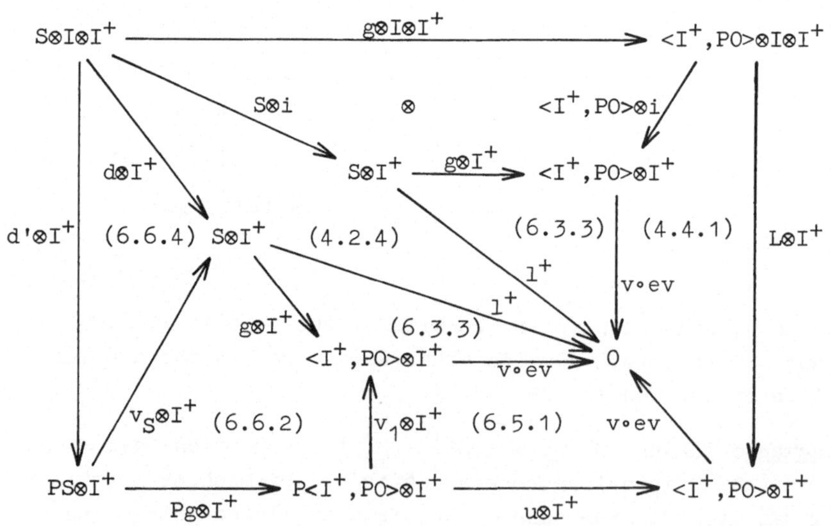

$$(6.6.5)$$

Vice versa given $d':S\otimes I \to PS$ satisfying (6.6.1) we define the automaton $A = (S,d,l)$ by

$$
(6.6.6) \qquad
\begin{aligned}
d &:= (S\otimes I \xrightarrow{\ d'\ } PS \xrightarrow{\ v_S\ } S) \qquad\text{and}\\[4pt]
l &:= (S\otimes I \xrightarrow{\ S\otimes i_1\ } S\otimes I^+ \xrightarrow{\ g\otimes I^+\ } <I^+,PO>\otimes I^+ \xrightarrow{\ ev\ } PO \xrightarrow{\ v\ } 0)
\end{aligned}
$$

We want to show $g = M(A)$. By definition of $M(A)$ in (6.3.3) this is equivalent to $v\circ ev\circ (g\otimes I^+) = l^+$. Since $S\otimes I^+$ is a co-product in $\underline{K}$ with injections $S\otimes i_n:S\otimes I^n \to S\otimes I^+$ $(n\in\mathbb{N})$ it suffices to show

$$
(6.6.7) \qquad (v\circ ev)\circ (g\otimes I^+)\circ (S\otimes i_n) = l^+\circ (S\otimes i_n) = l_n \qquad (n\in\mathbb{N})
$$

which will be verified by induction on $n\in\mathbb{N}$.

For $n = 1$ (6.6.7) is satisfied by definition of l .

Now we assume (6.6.7) to be valid for $n\in\mathbb{N}$ then the assertion for $n+1$ follows from the diagram (6.6.8) below.

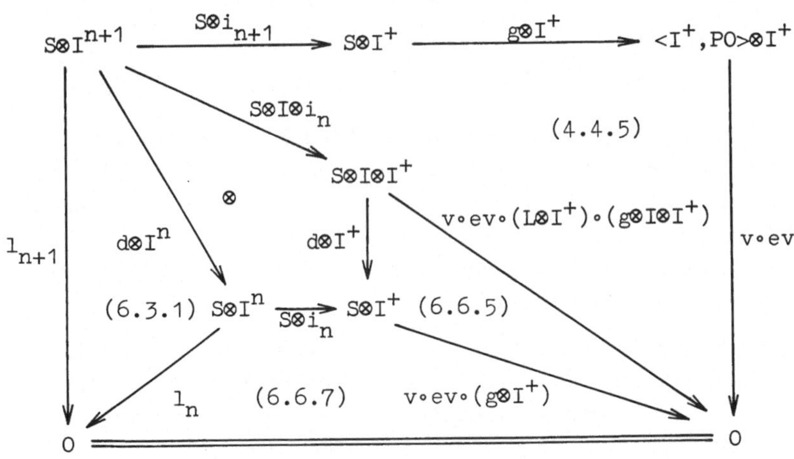

$$
(6.6.8) \qquad\qquad\qquad\qquad \blacksquare
$$

<u>6.7 Theorem</u> (B e h a v i o r   C h a r a c t e r i z a -
t i o n) : A $\underline{K}$'-morphism $m:B \to <I^+,PO>$ is the behavior of
an automaton A in the pseudoclosed category $(\underline{K},\otimes)$ if
  (i) m is a canonical representative of a morphism in $\mathfrak{M}$
      (cf. 4.6) and
(ii) there exists a $\underline{K}$'-morphism $d':B\otimes I \to PB$ satisfying:

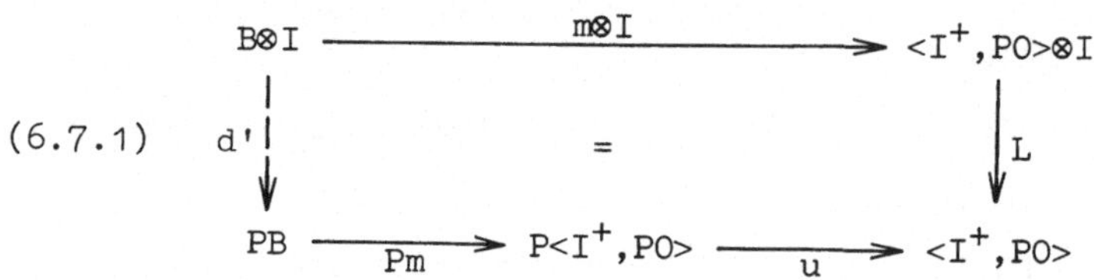

$$(6.7.1)$$

For the definition of L, u and Pm confer 6.6.
In this case an automaton A realizing the behavior m ,
i.e. $m(A) = m$, is given by $A = (B,d,l)$ with

$$(6.7.2) \quad d := (B\otimes I \xrightarrow{d'} PB \xrightarrow{v_B} B) \quad \text{and}$$

$$l := (B\otimes I \xrightarrow{B\otimes i_1} B\otimes I^+ \xrightarrow{m\otimes I^+} <I^+,PO>\otimes I^+ \xrightarrow{ev} PO \xrightarrow{v} O)$$

Vice versa given an automaton $A = (S,d,l)$ the behavior
$m(A):E(A) \to <I^+,PO>$ satisfies the conditions (i) and (ii)
provided that the $\underline{K}$'-morphism $e(A):S \to E(A)$ in the canonical $\mathfrak{E}$-$\mathfrak{M}$-factorization $m(A)\circ e(A) = M(A)$ (cf. (6.3.4)) is a
retraction in $\underline{K}$, i.e. there is a $\underline{K}$-morphism $c:E(A) \to S$
satisfying $e(A)\circ c = id_{E(A)}$.

<u>Proof</u>: By theorem 6.6 each $m:B \to <I^+,PO>$ in $\underline{K}$' satisfy-
ing the condition (ii) is the machine morphism of an automa-
ton A , i.e. $m = M(A)$. Using (6.6.6) A is explicitly
given by (6.7.2). Furthermore the condition (i) for m
implies $m = M(A) = m(A)$.
Vice versa given an automaton A it will be shown in 7.4
that there is an automaton A_c with machine morphism
$M(A_c) = m(A)$ provided that c is a coretraction of e(A) .

Now $m(A)$ satisfies condition (i) by construction of $m(A)$ in (6.3.4) and (ii) by theorem 6.6 because $M(A_c) = m(A)$ is the machine morphism of A_c .   ∎

<u>Interpretation</u>: According to our examples in 6.3 condition (i) asserts that B is a subset resp. subspace of $\langle I^+, PO \rangle$ with inclusion $m : B \to \langle I^+, PO \rangle$ which is closed under left shift up to union by condition (ii). In more detail we get the following interpretation of (6.7.1).

1. (N o n d e t e r m i n i s t i c   A u t o m a t a) :
For each $f \in B$ and $x \in I$ there is a non-empty subset $d'(f,x)$ of B such that the left shift $f \circ L_x$ is the union of all $f' \in d'(f,x)$ because we have by (6.7.1):

$$f \circ L_x = L \circ (m \times I)(f,x) = u \circ Pm \circ d'(f,x) = u(d'(f,x)) = \bigcup_{f' \in d'(f,x)} f'$$
(cf. 6.5).

Vice versa the surjective function $e(A) : S \to E(A)$ has clearly a coretraction $c : E(A) \to S$ which is a choice function assigning to each element $f \in E(A)$ an arbitrary $s \in e(A)^{-1}(f)$. Thus we have the following characterization: A subset $B \subseteq \langle I^+, P'O \rangle$ is the behavior of a nondeterministic automaton iff for each $f \in B$ and $x \in I$ the left shift $f \circ L_x$ can be represented as the union of elements of B .

2. (R e l a t i o n a l   A u t o m a t a) : Replacing $P'O$ by PO we get exactly the same behavior characterization as above.

3. (S t o c h a s t i c   A u t o m a t a) : For each $f \in B$ and $x \in I$ there is a probability distribution $F := d'(f,x) \in PB$ on B such that the left shift $f \circ L_x$ of the stochastic channel f is the sum of all $f' \in B$ weighted by $F(f')$:

$$f \circ L_x = L \circ (m \times I)(f,x) = u \circ Pm \circ d'(f,x) = u(d'(f,x)) = u(F)$$
$$= \sum_{f' \in B} F(f') \cdot f' \qquad \text{(cf. 6.5).}$$

The left shift $f \circ L_x$ is a stochastic channel $f \circ L_x : I^+ \to O$ defined by $f \circ L_x(w)(y) = f(xw)(y)$ for all $w \in I^+$, $y \in O$.
Since $e(A)$ was shown to have a coretraction in $(\underline{Set}, \times)$ we

obtain the following characterization:

A subset $B \subseteq \langle I^+, PO \rangle$ of all stochastic channels from I^+ to
O is the behavior of a stochastic automaton iff for each
f∈B and x∈I there is a probability distribution F on B
such that the left shift $f \circ L_x$ is the sum of all stochastic
channels in B weighted by F.

4. (R e l a t i o n a l   T o p o l o g i c a l   A u t o -
m a t a) : In this case the additional assumption on e(A)
to have a coretraction in $(\underline{K}, \otimes) = (\underline{RelTop}, \otimes)$ is not satisfied in general but only if e(A) is open in $(\underline{Top}, \otimes)$. In
other words we have to assume that for each open subset U of
S the set of all input-output relations $1^+(s,-):I^+ \to O$ is
open in E(A). In this case we get a continuous function
$d':E(A) \otimes I \to PE(A)$. The rest of the interpretation is similar
to examples 1 and 2. More details are given in [32].

<u>Remark</u> (F i n i t e   B e h a v i o r s) : Given a class
of finite objects $\underline{F}$ in $\underline{K}'$ as defined in the remark of
4.8 we can in the same way conclude: A behavior $m:S \to \langle I^+, PO \rangle$
satisfying (6.7.1) is realizable by an automaton $A = (S,d,1)$
with "finite" state object S∈$\underline{F}$ iff B is "finite", i.e.
B∈$\underline{F}$. This can be applied to characterize finite nondeterministic, relational and stochastic automata or compact
relational topological automata.

<u>6.8 Examples of Automata in Pseudoclosed Categories</u>:
In addition to the examples of nondeterministic type automata given in 1.12, which have already been discussed in 6.3,
we want to give a list of some other types of pseudoclosed
categories $(\underline{K}, \otimes)$ relative $(\underline{K}', \otimes)$ which satisfy the general assumptions given in 6.1 . The application of our
theory to the corresponding automata in $(\underline{K}, \otimes)$ is left to
the reader. Some explanations are given below.

1. (<u>ND</u>,×)   rel. (<u>Set</u>,×)   for nondeterministic automata
2. (<u>Rel</u>,×)   rel. (<u>Set</u>,×)   for relational automata
3. (<u>Stoch</u>,×)   rel. (<u>Set</u>,×)   for stochastic automata
4. (<u>RelTop</u>,⊗) rel. (<u>Top</u>,⊗)   for relational topological
                                automata
5. (<u>RelCG</u>,π)   rel. (<u>CG</u>,π)   for compactly generated
                                relational automata
6. (<u>RelMetr</u>,×) rel. (<u>Metr</u>,×) for relational metric automata
7. (<u>RelTol</u>,⊗) rel. (<u>Tol</u>,⊗)   for relational tolerance
                                automata
8. (<u>PDStoch</u>,×) rel. (<u>PD</u>,×)   for partial stochastic
                                automata
9. (<u>L-Rel</u>,×)   rel. (<u>Set</u>,×)   for L-relational automata

<u>Explanations</u>: 1. - 4. According to the examples in 6.2 the
categories (<u>ND</u>,×) of nondeterministic functions, (<u>Rel</u>,×)
of relations, and (<u>Stoch</u>,×) of stochastic channels are
pseudoclosed relative the closed category (<u>Set</u>,×) of sets,
and the category (<u>RelTop</u>,⊗) of lower semi-continuous
relations is pseudoclosed relative (<u>Top</u>,⊗) which is the
closed category of topological spaces with biproduct
(cf. 1.10,5). The interpretation of the theory for the corresponding types of automata is given in 6.3 and 6.7.

5. (<u>RelCG</u>,π) denotes the category of compactly generated
Hausdorff-spaces and continuous relations. A relation
$f : A \to B$ is called continuous if the inverse image of open
sets is open, of closed sets is closed and f is point-
compact, i.e. for each $x \in A$ the set $f(x) \subseteq B$ is supposed
to be compact. In [32] it is shown that (<u>RelCG</u>,π) is a
pseudoclosed category relative (<u>CG</u>,π) (cf. 4.9,5).

6. Based on the closed category (<u>Metr</u>,×) of metric spaces
(cf. 4.9,6) we can define the category (<u>RelMetr</u>,×) of
metric spaces and point-compact decreasing relations which
is pseudoclosed relative (<u>Metr</u>,×) . A relation $f : A \to B$ is
called decreasing if the corresponding function $f : A \to \wp B$
is decreasing with respect to the following definition of a

distance for non-empty compact subsets U and V of B :

$$d(U,V) := \min \{ \sup_{u \in U} \inf_{v \in V} d(u,v) + \sup_{v \in V} \inf_{u \in U} d(u,v), 1 \}$$

and $d(\phi,W) := \begin{cases} 0 & W = \phi \\ 1 & W \neq \phi \end{cases}$.

7. Given tolerance spaces (A,r) and (A',r') (cf. 4.9,7)
a tolerance relation is a relation f:A → A' such that x r y
implies f(x) r' f(y), meaning that for each x'∈f(x) there
exists y'∈f(y) with x' r' y' and vice versa. It is easy
to verify that the category (RelTol,⊗) of tolerance spaces
and tolerance relations is pseudoclosed relative (Tol,⊗)
as defined in 4.9,7.

8. Let (PDStoch,×) be the category of partially defined
stochastic channels f:A → B , i.e. partial functions
f:A → PB where PB is the set of all (discrete) probability distributions on B . Similar to example 3 it can be
shown that (PDStoch,×) is pseudoclosed relative (PD,×)
which is closed by 4.3,2.

9. Let L be a complete lattice with a multiplication
•:L×L → L preserving least upper and greatest lower bounds,
called complete lattice ordered semi-group in [42]. Then the
category L-Rel of L-relations consists of sets as objects
and functions f:A×B → L as morphisms f∈Mor L-Rel (A,B) .
For L = {0,1} f:A×B → L is the characteristic function of
a relation from A to B such that L-Rel coincides with
Rel in this case. In fact it can be also shown that L-Rel
is pseudoclosed relative (Set,×) .

6.9 Remark (G e n e r a l i z a t i o n s) : In analogy
to remark 4.10 for automata in closed categories we can
generalize our results to variable input- and output-objects
I resp. O taking arbitrary K'-morphisms in the O-compo-
nent but only K'-retractions f_I:I → I' . As remarked in
1.14 our theory can be reformulated for Moore-automata
replacing 1:S⊗I → O by m:S → O . This leads to a machine

morphism of the form $M(A):S \to \langle I^*,PO\rangle$ for Moore-automata A
in pseudoclosed categories. According to the remark in 6.2
there are similar approaches to this case introduced by
M. A. A r b i b   and   E. G. M a n e s   in [6] and by
E. B u r r o n i   in [19].

7. Reduction and Minimization of Automata in Pseudoclosed Categories

Automata in pseudoclosed categories, including nondeterministic, relational, stochastic and relational topological automata, have been introduced in chapter 6. In 2.7 it is shown that,in contrast to the deterministic case,reduction and minimization do   not coincide for nondeterministic automata. Thus we have to study reduction, minimization and realization problems for automata in pseudoclosed categories separately. The construction of an observable realization, already given in (6.7.2), turns out to be a weak minimal realization functor (cf. 7.8). Considering reduction and minimization we will construct for each automaton A in a pseudoclosed category an equivalent observable automaton A' and an equivalent reduced automaton R(A) together with a reduction u(A):A → R(A) (cf.7.3, 7.4). Uniqueness and other properties of these constructions are studied in 7.7 and 7.8 using the theory and classification of systematics which are introduced in chapter 3. Unfortunately, observable automata are not minimal in the sense of systematics in general but only with respect to "weak morphisms". This is the reason for the fact that equivalent observable automata are not necessarily isomorphic, but that only their state objects are. On the other hand minimality can be obtained regarding "strong observable" automata meaning that not only the states but also different subsets of states are inequivalent .

7.1 General Assumptions: According to 6.1 ($\underline{K}$,⊗) is a pseudoclosed category relative the closed category ($\underline{K}$',⊗) which is assumed to have countable coproducts and an $\mathfrak{E}$-$\mathfrak{M}$-factorization with canonical representatives in $\mathfrak{M}$. Moreover we will

use most of the constructions given in chapter 6 and the terminology of systematics in chapter 3. According to 1.11 re-
mark 3 $\underline{K}$-$\underline{K}$'-$\underline{Aut}$ will denote the category of automata in the
pseudoclosed category $(\underline{K},\otimes)$ with the additional property that
the automata morphisms belong to the subcategory $\underline{K}$' of $\underline{K}$.

For the construction of reduced automata in 7.3 we will as-
sume that $\underline{K}$' has "$\underline{\text{large cointersections}}$" meaning that for
each class $\tilde{I}$ and each family $f_i:S \to S_i$ $(i\in\tilde{I})$ of $\underline{K}$'-epi-
morphisms there is an object $\bar{S}$ together with a family of
$\underline{K}$'-morphisms $u_i:S_i \to \bar{S}$ $(i\in I)$, called injections, satisfying:

(a) $u_i \circ f_i = u_j \circ f_j$ for all $i,j\in\tilde{I}$ and

(b) for all other families $g_i:S_i \to S'$ of $\underline{K}$'-morphisms sat-
    isfying (a) there is a unique $\underline{K}$'-morphism $g:\bar{S} \to S'$ such
    that we have $g \circ u_i = g_i$ for all $i\in\tilde{I}$ in diagram (7.1.1):

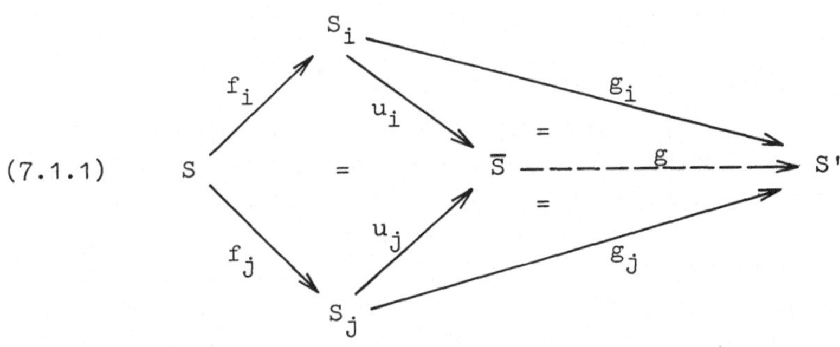

(7.1.1)

<u>Remarks</u>: We have used the word "large cointersections" in-
stead of "cointersections" because $\tilde{I}$ is allowed to be a
proper class and not only a set. If we have cointersections
in $\underline{K}$' , i.e. $\tilde{I}$ is always a set, we also have large cointersections provided that $\underline{K}$' is $\underline{\text{cowell-powered}}$ in the
following sense:
For each object S in $\underline{K}$' there is a set $\tilde{I}$ and a family
of epimorphisms $f_i:S \to S_i$ $(i\in\tilde{I})$ such that for each epimorphism $f:S \to S'$ there is an $i_o\in\tilde{I}$ and an isomorphism

$h:S_{i_o} \tilde{\to} S'$ satisfying $h \cdot f_{i_o} = f$. In other words: For each S in $\underline{K}'$ there is only a set of "non-equivalent" epimorphisms with the same domain S.

<u>Interpretation</u>: The notion of a cointersection is dual to the well known construction of intersections of subsets or subspaces. In the category <u>Set</u> of sets the cointersection of a family of surjective functions $f_i:S \to S_i$ $(i \in \tilde{I})$ is a quotient set

$$\bar{S} = \bigcup_{i \in \tilde{I}} S_i \Big/ \bar{R}$$

of the disjoint union of all the sets S_i $(i \in \tilde{I})$ where $\bar{R}$ is the equivalence relation generated by the relation R on the disjoint union defined by $s_i R s_j$ for $s_i \in S_i$ and $s_j \in S_j$ iff there exists an $s \in S$ satisfying $f_i(s) = s_i$ and $f_j(s) = s_j$. In the category <u>Top</u> of topological spaces we use the same construction for $\bar{S}$ and take the final topology with respect to all the functions $u_i:S_i \to \bar{S}$ $(i \in \tilde{I})$ which assign to each $s_i \in S_i$ the corresponding class $[s_i] \in \bar{S}$.

To show that <u>Set</u> and <u>Top</u> have also large cointersections it suffices to show that they are cowell-powered. For each set S the class of all natural mappings $nat_i:S \to S_i$ $(i \in \tilde{I})$ to all the quotient sets S_i of S is clearly a set and for each surjective function $f:S \to S'$ there is a unique isomorphism $h:\bar{S} \to S'$ satisfying $h \cdot nat = f$ where $\bar{S}$ is the quotient set of S caused by f and $nat:S \to \bar{S}$ the corresponding natural mapping. A similar argument is true in <u>Top</u> replacing each S_i by a family S_{ij} with $j \in J_i$ ranging over a subset of all topologies on S_i .

Before we start with the construction of reduced automata let us reformulate the basic definitions for automata in pseudoclosed categories (cf. 5.4) :

<u>7.2 Definitions</u>: 1. An automaton A in $(\underline{K}, \otimes)$ is called <u>observable</u> (or $\mathfrak{M}$-<u>minimal</u>) if the machine morphism $M(A)$ of A (cf. (6.3.3)) belongs to the class $\mathfrak{M}$.

2. An <u>automata morphism</u> $f:A \to A'$ in <u>K</u>-<u>K</u>'-<u>Aut</u> is a
<u>K</u>'-morphism $f:S \to S'$ satisfying

(7.2.1) $d' \circ (f \otimes I) = f \circ d$ and $l' \circ (f \otimes I) = l$

and it is a <u>reduction</u> if f belongs to the class $\mathfrak{C}$.

3. According to 3.2 A is called <u>reduced</u> if each reduction
$f:A \to A'$ to an arbitrary automaton A' in $(\underline{K},\otimes)$ is already an isomorphism.

4. Two automata A and A' are called <u>equivalent</u> if they
have the same behavior $E(A) = E(A')$, or more precisely
$m(A) = m(A')$ (cf. (6.3.4)).

<u>Remark</u>: For each reduction $f:A \to A'$ we already have
$E(A) = E(A')$ which is assumed in 3.2 . This can be verified
in the same way as in the remark of 5.4 using the property
$M(A') \circ f = M(A)$ which will be shown in (7.6.5).

As motivated above, the categories <u>Set</u> and <u>Top</u> have
large cointersections such that the following construction
of reduced automata can be applied to nondeterministic,
relational, stochastic and relational topological automata
(cf. 6.3).

<u>7.3 Proposition</u> (R e d u c t i o n) : For each automaton
A in <u>K</u>-<u>K</u>'-<u>Aut</u> there is a reduced automaton $R(A)$ and a
reduction $u(A):A \to R(A)$ provided that the category <u>K</u>'
has large cointersections (cf. 7.1).

<u>Proof</u>: Let $A = (S,d,l)$ be an automaton in <u>K</u>-<u>K</u>'-<u>Aut</u> . The
reduced automaton $R(A)$ and the reduction $u(A):A \to R(A)$
will be constructed by a large cointersection ranging over
all reductions $f:A \to A'$ with fixed domain A . More precisely we will construct the cointersection of the corresponding
<u>K</u>'-morphisms $f:S \to S'$ in the category <u>K</u>'. All these mor-
phisms f which belong to the class $\mathfrak{C}$ constitute a family
$f_i:S \to S_i$ $(i \in \tilde{I})$ where $\tilde{I}$ is a proper class in general. Let $\bar{S}$
be the large cointersection of this family with injections

$u_i : S_i \to \bar{S}$ $(i \in \tilde{I})$ and $u(A)$ defined by the composition $u(A) = u_i \circ f_i$ for all $i \in \tilde{I}$.

(7.3.1)

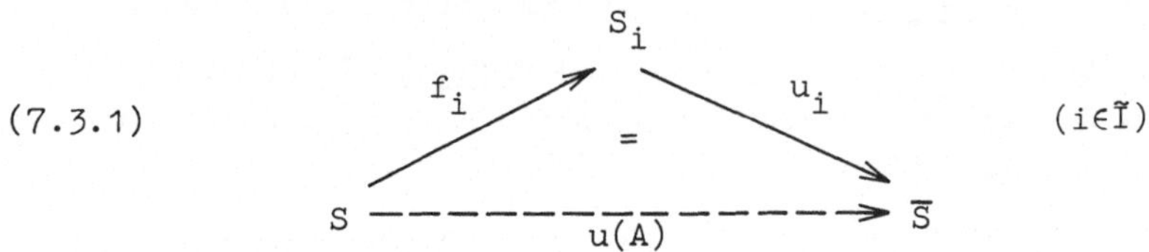

Since $J \circ (-\otimes I) : \underline{K}' \to \underline{K}$ is a left adjoint functor by the lemma in 6.2 and left adjoint functors preserve colimits and especially cointersections (cf. 12.10) the object $\bar{S} \otimes I$ together with the morphisms $u_i \otimes I : S_i \otimes I \to \bar{S} \otimes I$ $(i \in \tilde{I})$ is a cointersection of the family $f_i \otimes I : S \otimes I \to S_i \otimes I$ $(i \in \tilde{I})$ in the category $\underline{K}$. Using the universal properties of the cointersection it will be possible to construct $\underline{K}$-morphisms $\bar{d} : \bar{S} \otimes I \to \bar{S}$ and $\bar{I} : \bar{S} \otimes I \to 0$ such that $\bar{A} = (\bar{S}, \bar{d}, \bar{I})$ is a reduced automaton and $u(A) : A \to \bar{A}$ is a reduction.

In the following diagram (7.3.2) the triangles are commutative by (7.3.1), as well as the squares (1) and (2) because $f_i : A \to A_i$ $(i \in \tilde{I})$ are automata morphisms.

(7.3.2)

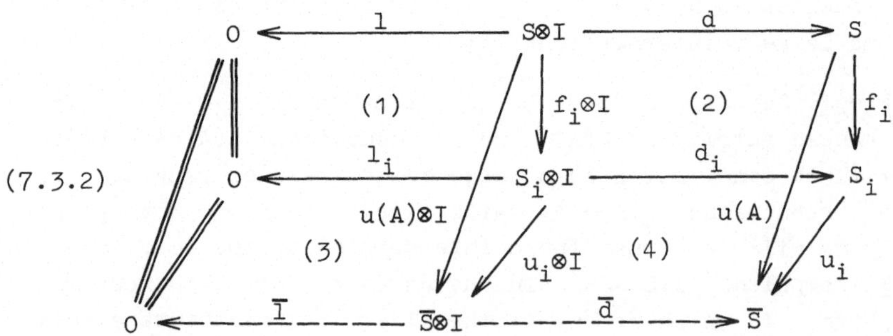

For the family $u_i \circ d_i$ $(i \in \tilde{I})$ we have

$$u_i \circ d_i \circ (f_i \otimes I) = u_i \circ f_i \circ d = u(A) \circ d = u_j \circ d_j \circ (f_j \otimes I)$$

for all $i, j \in \tilde{I}$. Hence there is a unique $\underline{K}$-morphism
$\bar{d}: \bar{S} \otimes I \to \bar{S}$ such that (4) is commutative for all $i \in \tilde{I}$
(cf. (7.1.1)). In the same way we find a unique I in (3)
of (7.3.2). Combining the subdiagrams in (7.3.2) we see that
$u(A): S \to \bar{S}$ is an automata morphism $u(A): A \to \bar{A}$ which is a
reduction because we will show $u(A) \in \mathfrak{E}$:
Let $m \circ e = u(A)$ be an $\mathfrak{E}$-$\mathfrak{M}$-factorization of $u(A)$ in (7.3.3).
Since $f_i \in \mathfrak{E}$ and $u_i \circ f_i = u(A) = m \circ e$ for all $i \in \tilde{I}$ by (7.3.1)
we have unique diagonal morphisms g_i in (7.3.3) satisfying
$g_i \circ f_i = e$ and $m \circ g_i = u_i$ (cf. 4.7,1). Hence, using the universal properties (7.1.1) of the cointersection we get a
unique g in (7.3.3) satisfying $g \circ u_i = g_i$ for all $i \in \tilde{I}$.
Now we have $m \circ g \circ u_i = m \circ g_i = u_i$ for all $i \in \tilde{I}$ which implies
$m \circ g = id_{\bar{S}}$. Thus $m \in \mathfrak{M}$ is a retraction and hence an isomorphism by 4.7,4 such that $u(A) = m \circ e$ belongs to $\mathfrak{E}$.

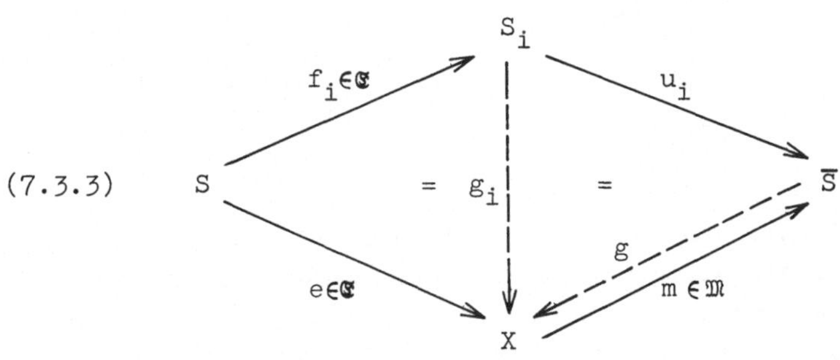

In order to show that $R(A) := \bar{A}$ is reduced we take an arbitrary reduction $f': \bar{A} \to A'$ and we will show that f' is
already an isomorphism. Since $u(A): A \to \bar{A}$ is a reduction
the same is true for $f' \circ u(A): A \to A'$. Thus we have $f' \circ u(A) =$
$= f_{i_o}$ for a suitable $i_o \in \tilde{I}$ by definition of the family
$[f_i]_{i \in \tilde{I}}$ which implies $u_{i_o} \circ f' \circ u(A) = u_{i_o} \circ f_{i_o} = u(A)$ and hence

$u_{i_0} \circ f' = id_S$ because $u(A)$ is an epimorphism. Thus $f' \in \mathfrak{C}$ is a coretraction and hence an isomorphism in $\underline{K}'$ by 4.7,4 , and also in $\underline{K}$-$\underline{K}'$-$\underline{Aut}$. ∎

<u>Example</u>: Consider the nondeterministic automaton A_6 defined by (7.3.4).

(7.3.4)

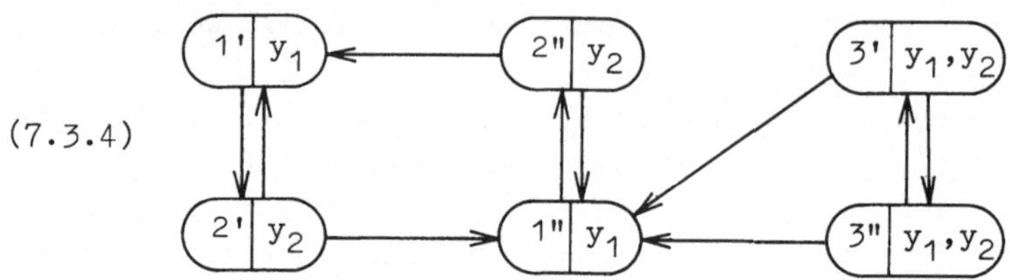

It is easy to see that the states i' and i'' for $i = 1,2,3$ have the same input-output behavior and also the same state transition with respect to equivalence of states. Hence we get a reduction $r:A_6 \to A_4$ to A_4 given by (7.3.5) (cf. (2.7.4))

(7.3.5)

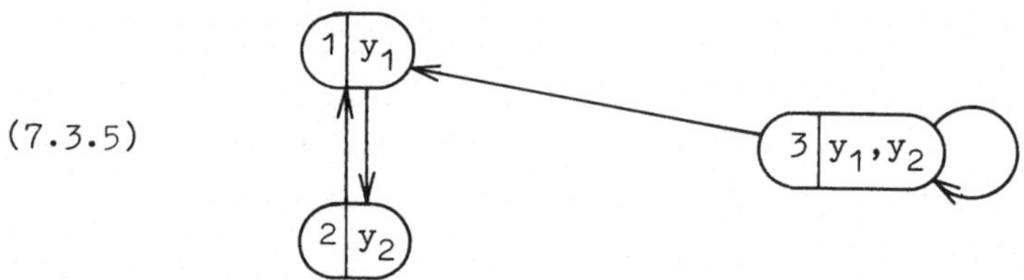

and r identifies the states i' and i'' to i for $i = 1,2,3$. All states of A_4 are pairwise inequivalent such that the automaton A_4 is reduced. Thus the theorem above is satisfied for A_6 defining $R(A_6) := A_4$ and $u(A_6) := r$. Another simple example for the construction of reduced automata is given in (2.7.3).

<u>7.4 Proposition</u> (M i n i m i z a t i o n) : Given an automaton A in $\underline{K}$-$\underline{K}'$-$\underline{Aut}$ and a canonical $\mathfrak{C}$-$\mathfrak{M}$-factorization $m(A) \circ e(A) = M(A)$ of the machine morphism $M(A)$ we assume that $e(A):S \to E(A) \in \mathfrak{C}$ has a coretraction $c:E(A) \to S$ in $\underline{K}$, i.e. $e(A) \circ c = id_{E(A)}$. Then there is an observable automaton

$A_c = (E(A), d_c, l_c)$ defined by the diagram (7.4.1)

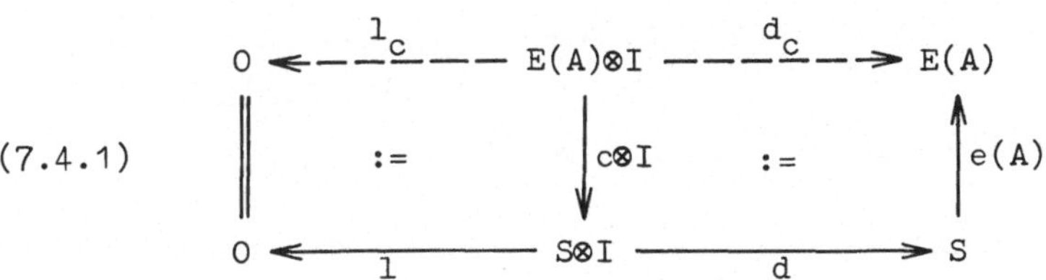

which is equivalent to A, especially we have

$$(7.4.2) \qquad M(A_c) = m(A) : E(A) \to \,<I^+, PO>$$

Remark: Different coretractions c_1 and c_2 of $e(A)$ lead to non-isomorphic observable automata A_{c_1} and A_{c_2} in general and both of them are equivalent to A (cf. (2.7.4), (2.7.5)). According to the interpretation in 6.7 the existence of a coretraction is obvious in the case $\underline{K}' = \underline{Set}$ taking a choice function c of the surjective function $e(A)$ (cf. 6.7,1). Hence the coretraction condition is satisfied for nondeterministic, relational and stochastic automata. But in the case $\underline{K}' = \underline{Top}$ $e(A)$ has to be an open function such that the inverse relation c is lower semi-continuous and this is an additional assumption for a relational topological automaton A (cf. [32]).

Proof: It suffices to show (7.4.2) because $M(A_c) = m(A) \in \mathfrak{M}$ implies that A_c is observable. Furthermore we have $m(A_c) = M(A_c) = m(A)$ using that $m(A) \circ id_{E(A)} = M(A_c)$ is a canonical $\mathfrak{E}\text{-}\mathfrak{M}$-factorization of $M(A_c)$. In order to verify (7.4.2) it remains to show

$$(7.4.3) \qquad 1_c^+ = 1^+ \circ (c \otimes I^+)$$

because (7.4.3) implies by (6.3.3)

$$v \circ ev \circ ((M(A) \circ c) \otimes I^+) = v \circ ev \circ (M(A) \otimes I^+) \circ (c \otimes I^+) = 1^+ \circ (c \otimes I^+)$$

$$= 1_c^+ = v \circ ev \circ (M(A_c) \otimes I^+) \quad .$$

Furthermore using the uniqueness properties of $v \circ ev$
(cf. (6.2.2)) we get $M(A) \circ c = M(A_c)$ and by $e(A) \circ c = id_{E(A)}$

$$(7.4.4) \qquad \frac{M(A_c) \circ e(A) = M(A) \circ c \circ e(A) = m(A) \circ e(A) \circ c \circ e(A)}{= m(A) \circ id_{E(A)} \circ e(A) = \underline{m(A) \circ e(A)} = M(A).}$$

This finally implies $M(A_c) = m(A)$ because $e(A)$ is an epi-morphism.

Hence it remains to verify (7.4.3). Using the coproduct properties of $E(A) \otimes i_n : E(A) \otimes I^n \to E(A) \otimes I^+$ condition (7.4.3) is equivalent to

$$1_c^+ \circ (E(A) \otimes i_n) = 1^+ \circ (c \otimes I^+) \circ (E(A) \otimes i_n) \quad \text{for all } n \in \mathbb{N}$$

and hence to

$$(7.4.5). \qquad (1_c)_n = 1_n \circ (c \otimes I^n) \qquad \text{for all } n \in \mathbb{N}$$

because of $(c \otimes I^+) \circ (E(A) \otimes i_n) = (S \otimes i_n) \circ (c \otimes I^n)$ and (6.3.2). Condition (7.4.5) is satisfied for n=1 by (7.4.1) and for n+1 with $n \geq 1$ we have by induction

$$\begin{aligned}
(1_c)_{n+1} &= (1_c)_n \circ (d_c \otimes I^n) & (6.3.1) \\
&= 1_n \circ (c \otimes I^n) \circ (d_c \otimes I^n) & \text{(induction hypothesis)} \\
&= 1_n \circ (c \otimes I^n) \circ (e(A) \otimes I^n) \circ (d \otimes I^n) \circ (c \otimes I \otimes I^n) & (7.4.1) \\
&= 1_n \circ (d \otimes I^n) \circ (c \otimes I^{n+1}) & \text{(see (7.4.6) below)} \\
&= 1_{n+1} \circ (c \otimes I^{n+1}) & (6.3.1)
\end{aligned}$$

In order to show the last but one step let us observe that $M(A) = M(A) \circ c \circ e(A)$ in (7.4.4) implies $1^+ = 1^+ \circ ((c \circ e(A)) \otimes I^+)$ by (6.3.3) and hence

$$(7.4.6) \qquad 1_n = 1_n \circ (c \otimes I^n) \circ (e(A) \otimes I^n)$$

by composition with $(S \otimes i_n)$ using (6.3.2). $\blacksquare$

An example for the minimization construction is given in 2.7.

<u>7.5 Remark</u> (R e a l i z a t i o n) : The behavior charac-terization in theorem 6.7 leads to the following realization construction: Given a behavior $m : B \to \langle I^+, PO \rangle$ which is a canonical representative of a morphism in $\mathfrak{M}$ and satisfies

(6.7.1), meaning that B is closed under left shift L up
to union, then the following automaton $A = (B,d,l)$ defined
in (6.7.2) and in (7.5.1) realizes the behavior $m: B \to <I^+, PO>$,
i.e. $m(A) = m$:

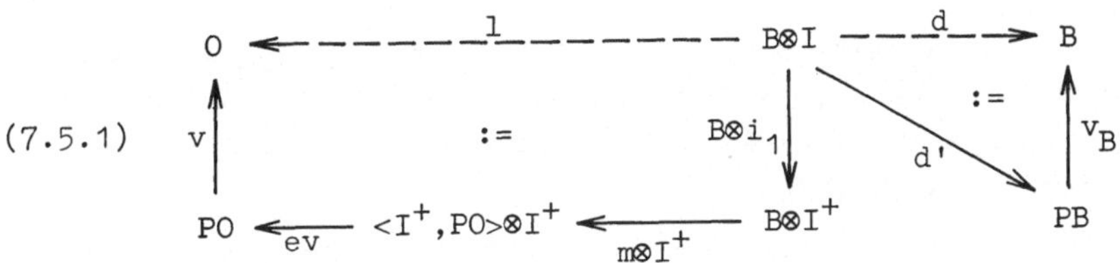

But let us point out again that the condition (6.7.1) for
the behavior $m(A): E(A) \to <I^+, PO>$ is not necessary in general unless $e(A)$ has a coretraction in $\underline{K}$. According to the
remark in 7.4 this is always the case for nondeterministic,
relational and stochastic but not for relational topological
automata in general.

7.6 Systematic of Automata in Pseudoclosed Categories:
In order to study the constructions of reduction and minimization in 7.3 and 7.4 systematically we consider the systematic of automata in pseudoclosed categories in the sense
of chapter 3 hoping to find nice properties of the subsys-
tematics defind by all reduced resp. observable automata.
According to 3.1 the system category $\underline{S}$ is the category
$\underline{K}$-$\underline{K}$'-$\underline{Aut}$ of automata in $(\underline{K}, \otimes)$ with morphisms in $\underline{K}'$ (cf. 7.1).
The behavior category $\underline{B}$ has as objects the behaviors
$m(A): E(A) \to <I^+, PO> \in \mathfrak{M}$ for all A in $\underline{K}$-$\underline{K}$'-$\underline{Aut}$ (cf. (6.3.4))
which are characterized in 6.7. Morphisms in $\underline{B}$ are $\underline{K}$'-mor-
phisms $g: E(A) \to E(A')$ satisfying $m(A) = m(A') \circ g$. Since
$m(A') \in \mathfrak{M}$ is a monomorphism there is at most one morphism in
$\underline{B}$ from $m(A)$ to $m(A')$. Thus $\underline{B}$ is a partially ordered
class using in addition the fact that each isomorphism in $\underline{B}$
is already an identity. Note that in this case $m(A)$ and

m(A') are both canonical representatives of the same equivalence class and hence equal.

It remains to show that the behavior construction in (6.3.4)
can be extended to a functor

$$E:\underline{K}\text{-}\underline{K}'\text{-}\underline{Aut} \to \underline{B} \ .$$

Given an automata morphism $f:A \to \tilde{A}$ in $\underline{K}\text{-}\underline{K}'\text{-}\underline{Aut}$ we have to
show that there is a $\underline{K}'$-morphism $E(f):E(A) \to E(\tilde{A})$ satis-
fying $m(\tilde{A})\circ E(f) = m(A)$. By definition f is a $\underline{K}'$-morphism
$f:S \to \tilde{S}$ satisfying

$$(7.6.1) \qquad \tilde{d}\circ(f\otimes I) = f\circ d \quad\text{and}\quad \tilde{I}\circ(f\otimes I) = 1 \ .$$

By induction it is easy to show

$$(7.6.2) \qquad \tilde{I}_n\circ(f\otimes I^n) = 1_n \qquad\text{for all}\quad n\in\mathbb{N}$$

which implies by (6.3.2) and the coproduct properties of
$S\otimes I^+$:

$$(7.6.3) \qquad \tilde{I}^+\circ(f\otimes I^+) = 1^+.$$

Since $f\in\underline{K}'$ we get by uniqueness of $M(A)$ in (6.3.3)

$$(7.6.4) \qquad M(\tilde{A})\circ f = M(A) \ .$$

Finally $E(f)$ is the unique diagonal morphism in (7.6.5)
which exists by lemma 4.7,1 because $e(A)\in\mathscr{E}$ and $m(A)\in\mathfrak{M}$.

Since $\underline{B}$ is a partially ordered class we have shown that
$E:\underline{K}\text{-}\underline{K}'\text{-}\underline{Aut} \to \underline{B}$ is a functor.

Sometimes it is useful to have also a weaker notion of automata morphisms given by the property (7.6.4). Especially the
morphism $e(A):S \to E(A)$ in the proposition 7.4 is no automata morphism in general but only a weak one by (7.4.4).

<u>Definitions</u>: The functor $E:\underline{K}\text{-}\underline{K}'\text{-}\underline{Aut} \to \underline{B}$, constructed
above, is called <u>behavior functor</u> and defines the <u>systematic</u>
$\underline{\underline{K}}\text{-}\underline{\underline{K}}'\text{-}\underline{\underline{Aut}} = (\underline{K}\text{-}\underline{K}'\text{-}\underline{Aut},\underline{B},E)$ <u>of automata in the pseudoclosed</u>
<u>category</u> $(\underline{K},\otimes)$ relative $(\underline{K}',\otimes)$.

Given two automata A , $\tilde{A}$ in $(\underline{K},\otimes)$ a $\underline{K}'$-morphism $f:S \to \tilde{S}$
is called <u>weak morphism</u> if f satisfies (7.6.4).

Let $\underline{K}$-$\underline{K}'$-$\underline{Aut}^W$ be the category of automata in $(\underline{K},\otimes)$ with weak morphisms and E^W the extension of E to $\underline{K}$-$\underline{K}'$-$\underline{Aut}^W$. E^W is equal to E on objects and defined on morphisms by (7.6.5) again. Then $\underline{\underline{K}}$-$\underline{\underline{K}}'$-$\underline{\underline{Aut}}^W = (\underline{K}$-$\underline{K}'$-$\underline{Aut}^W,\underline{B},E^W)$ is called weak systematic of automata in $(\underline{K},\otimes)$.

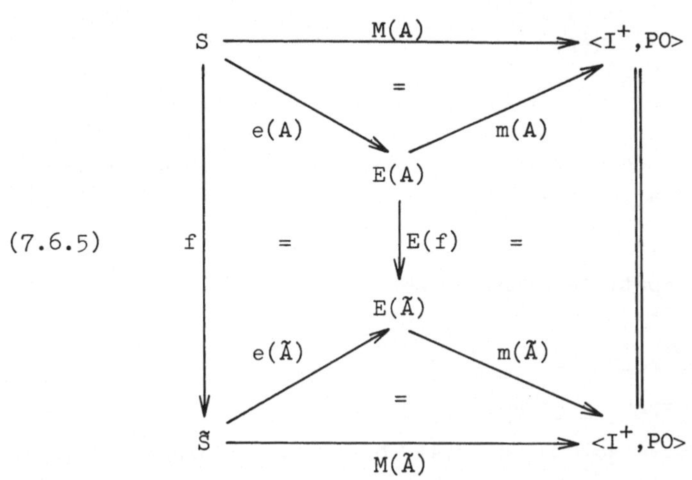

(7.6.5)

Now we come to the central theorems of reduction and minimization of automata in pseudoclosed categories.

7.7 Theorem (R e d u c t i o n) : The subsystematic $\underline{\underline{K}}$-$\underline{\underline{K}}'$-$\underline{\underline{Aut}}_{red}$ of $\underline{\underline{K}}$-$\underline{\underline{K}}'$-$\underline{\underline{Aut}}$ defined by all reduced automata in $(\underline{K},\otimes)$ is a reduced and realizing subsystematic in the sense of 3.3 provided that $\underline{K}'$ has large cointersections (cf. 7.1) and pushouts (cf. 12.9). In more detail we have by 3.3 and 3.4:

1. For each automaton A in $\underline{K}$-$\underline{K}'$-$\underline{Aut}$ there is an equivalent reduced automaton $R(A)$ and a reduction $u(A):A \to R(A)$ which is constructed by a large cointersection (cf. (7.3.1), (7.3.2)).

2. Given a reduced system A' in $\underline{K}$-$\underline{K}'$-$\underline{Aut}_{red}$ and an automata morphism $f:A \to A'$ in $\underline{K}$-$\underline{K}'$-$\underline{Aut}$ there is a unique mor-

phism $f':R(A) \to A'$ in (7.7.1) satisfying $f' \circ u(A) = f$.

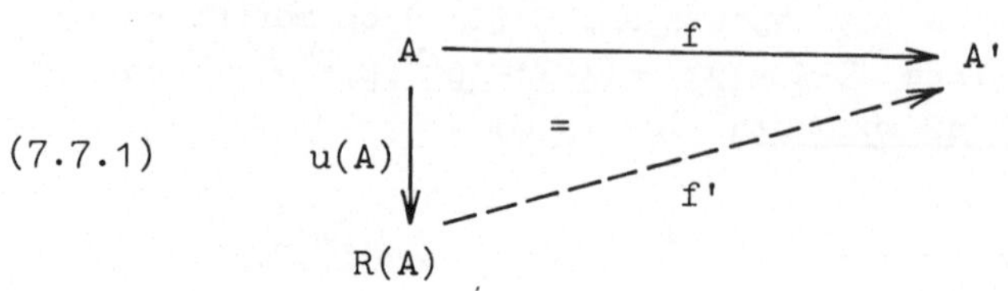

(7.7.1)

3. Moreover the construction $R(A)$ of reduced automata can be extended to a reduction functor

$$R:\underline{K}\text{-}\underline{K}'\text{-}\underline{Aut} \to \underline{K}\text{-}\underline{K}'\text{-}\underline{Aut}_{red}$$

which is compatible with the behavior, i.e.

$$E' \circ R = E$$

where E' is the restriction of the behavior functor E to the subcategory $\underline{K}\text{-}\underline{K}'\text{-}\underline{Aut}_{red}$ of $\underline{K}\text{-}\underline{K}'\text{-}\underline{Aut}$.

4. The reduced automaton $R(A)$ of A is uniquely determined up to isomorphism by the property that there is a reduction $u(A):A \to R(A)$ from A to a reduced automaton.

5. For each reduction $f:A \to A_1$ there is a unique reduction $f_1:A_1 \to R(A)$ in (7.7.2) satisfying $f_1 \circ f = u(A)$,

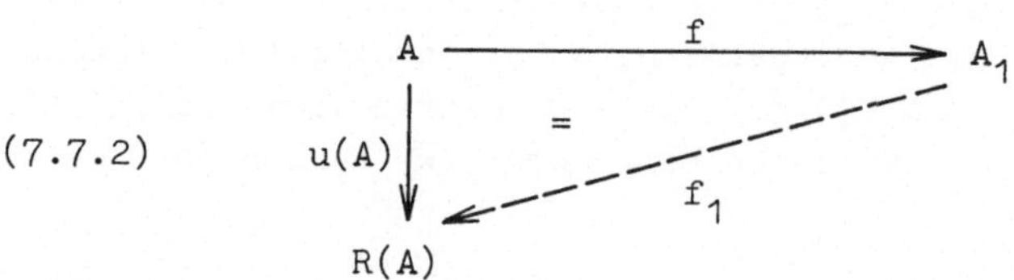

(7.7.2)

meaning that the reduction process is decomposable.

6. Given automata A_1 and A_2 the reduced automata $R(A_1)$ and $R(A_2)$ are isomorphic iff A_1 and A_2 are R-equivalent, i.e.

there is a possibly alternating chain of reductions between A_1 and A_2 , e.g.

$$A_1 \longleftarrow A_3 \longrightarrow A_4 \longleftarrow A_2 \ .$$

Moreover R-equivalence implies equivalence but in general not vice versa.

<u>Remark</u>: In categorical terms assertion 2 means that $\underline{K}$-$\underline{K}$'-$\underline{Aut}_{red}$ is a reflexive subcategory of $\underline{K}$-$\underline{K}$'-$\underline{Aut}$ and the corresponding "special problem" is stated in 5. (cf. [23,68]).
The interpretation of the cointersection construction of $R(A)$ is dicussed in 7.1. The pushout of two morphisms $[f_i : K_i \to L]_{i=1,2}$ has exactly the same universal property as the cointersection of the f_i but the f_i are not assumed to be epimorphisms (cf. 12.9). In all our examples in 6.3 pushouts can be constructed similar to cointersections (cf. interpretation in 7.1) which implies that the theorem is applicable to all these examples.

<u>Proof</u>: Using proposition 7.3 it remains to verify assertion 2 in order to show that $\underline{K}$-$\underline{K}$'-$\underline{Aut}_{red}$ is a reduced subsystematic (cf. 3.3). Assertions 3 to 6 are clear by theorem 3.4, 1-4 and 3.4,5 shows that $\underline{K}$-$\underline{K}$'-$\underline{Aut}_{red}$ is realizing because $\underline{K}$-$\underline{K}$'-$\underline{Aut}$ is realizing by definition of $\underline{B}$ and E in 7.6. In order to verify assertion 2 let $f:A \to A'$ be an automata morphism from A to a reduced automaton A'. Now we construct the pushout of the $\underline{K}$'-morphisms $f:S \to S'$ and $u(A):S \to \bar{S} \in \mathcal{C}$ leading to diagram (7.7.3) which corresponds to (7.3.1).

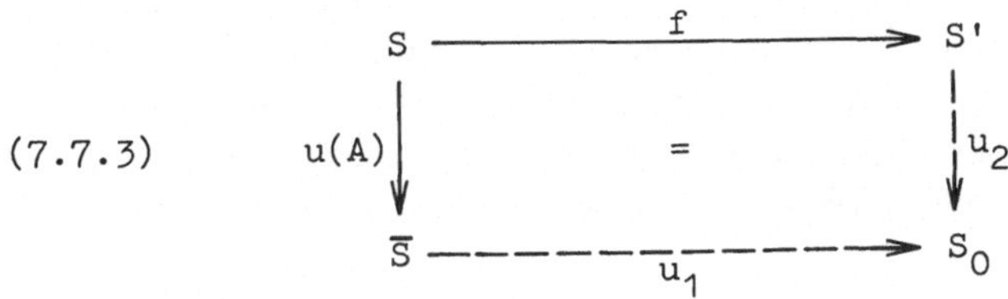

But note that $u(A)$ in $(7.7.3)$ is already given. A construction similar to $(7.3.2)$ shows that there is a unique automaton A_o with state object S_o such that u_1 and u_2 become automata morphisms $u_1 : R(A) \to A_o$, $u_2 : A' \to A_o$ respectively. Similar to $(7.3.3)$ $u(A) \in \mathfrak{C}$ implies $u_2 \in \mathfrak{C}$ such that u_2 is a reduction, and hence an isomorphism in $\underline{K}\text{-}\underline{K}'\text{-}\underline{Aut}$ because A' is reduced by assumption. Defining f' in $(7.7.1)$ by $f' := u_2^{-1} \circ u_1$ we have $f' \circ u(A) = u_2^{-1} \circ u_1 \circ u(A) = u_2^{-1} \circ u_2 \circ f = f$ and f' is uniquely determined because $u(A)$ is an epimorphism. Finally let us note that $(7.7.2)$ is a direct consequence of the constructions in $(7.3.1)$ and $(7.3.2)$ because we have $f = f_{i_o}$ for suitable $i_o \in I$ without loss of generality such that f_1 can be defined by $f_1 := u_{i_o}$. ∎

<u>Motivation</u>: As we have seen above there is a reduction $u(A) : A \to R(A)$ from A to the reduced automaton $R(A)$ but $R(A)$ is not observable in general (cf. 2.7). On the other hand using proposition 7.4 there is an observable automaton A_c equivalent to A but in general only a weak automata morphism $e(A) : A \to A_c$. This motivates to regard minimization with respect to the weak systematic $\underline{K}\text{-}\underline{K}'\text{-}\underline{Aut}^W$ of automata in $(\underline{K}, \otimes)$ (cf. 7.6).

<u>7.8 Theorem</u> (M i n i m i z a t i o n) : The subsystematic $\underline{K}\text{-}\underline{K}'\text{-}\underline{Aut}^W_{obs}$ of the weak systematic $\underline{K}\text{-}\underline{K}'\text{-}\underline{Aut}^W$ defined by all observable automata in $(\underline{K}, \otimes)$ is minimal, reduced and realizing in the sense of 3.3. We only assume that each morphism $e \in \mathfrak{C}$ has a coretraction c in $\underline{K}$.

<u>Remark</u>: Of course the notions "minimal", "reduced" and "realizing" defined in 3.3 have now to be regarded with respect to weak automata morphisms. Only some interesting applications of the theorems in 3.4 and 3.6 and other important properties of observable automata are stated now in more detail:

1. For each automaton A in the pseudoclosed category $(\underline{K}, \otimes)$ there is an equivalent observable automaton A_c and a weak

automata morphism $e(A):A \to A_c$. The explicit construction
of A_c , depending on the choice of the coretraction c of
$e(A)$, is given in (7.4.1).

2. An automaton A' is observable iff it is minimal with
respect to $\underline{K}-\underline{K}'-\underline{\underline{Aut}}^W$, i.e. for each automaton A satisfying
$E(A) \subseteq E(A')$ there is exactly one weak morphism $f:A \to A'$
which belongs to $\mathfrak{C}$ in the case $E(A) = E(A')$.

3. Equivalent observable automata are isomorphic with respect to weak morphisms, esɪ ɜcially they have isomorphic
state objects but they are not isomorphic in general (cf.
2.7).

4. Given an automaton A in $(\underline{K},\otimes)$ there is a reduction
$f:A \to A'$ from A to an observable automaton iff A satisfies the "<u>condition of representatives</u>" meaning that there
is a coretraction c of $e(A)$ such that the diagram
(7.8.1) commutes:

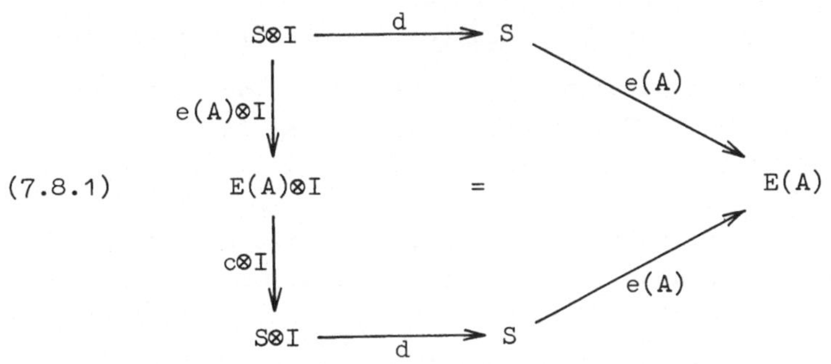

5. Each observable automaton is reduced. But a reduced automaton A is observable iff A satisfies the condition of
representatives.

6. Given a cardinality function in the sense of 3.7 an automaton A' with finite cardinality is observable iff we have
$card(A') \leqq card(A)$ for all automata A equivalent to A'.

134

<u>Interpretation</u>: The assumption that each $e \in \mathcal{C}$ has a core-
traction is already interpreted in 6.7 and 7.4 such that the
theorem is applicable to nondeterministic, relational and
stochastic automata but only to those relational topological
automata A for which e(A) is an open function.
The existence of a weak morphism $f:A \to A'$ means that for
each state $s \in S$ there is an equivalent state $s' \in S'$, i.e.
$M(A)(s) = M(A')(s')$, but there is no compatibility with the
state transition functions d and d' in general. $e(A):A \to A_c$
is such a weak morphism but e(A) is compatible with the
state transition iff A satisfies the condition of representatives in (7.8.1). In the case of nondeterministic automata (7.8.1) means the following: Given a pair of equivalent
states $s,s' \in S$ for each $x \in I$ and next state $s_1 \in d(s,x)$ of
s there is an equivalent next state $s_1' \in d(s',x)$ of s'
and vice versa. Note that $s' \in c \circ e(A)(s)$ implies that s'
is equivalent to s because we have:

$$M(A)(s) = m(A) \circ e(A)(s) = m(A) \circ e(A) \circ c \circ e(A)(s)$$
$$= M(A) \circ c \circ e(A)(s) = M(A)(s') \quad \text{for all } s' \in c \circ e(A)(s)$$

An example of an automaton which satisfies the condition of
representatives (7.8.1) is the automaton A_6 in (7.3.4). Due
to assertion 4 the reduced automaton $R(A_6)$ is observable in
this case.
Finally application of assertion 6 to finite nondeterministic, relational and stochastic automata means that an automaton is observable iff it has a minimal number of states.

<u>Proof of the theorem</u>: By proposition 7.4 for each automaton
A there is an equivalent observable automaton A_c such
that by definition of the behavior category $\underline{B}$ in 7.6 the
subsystematic $\underline{\underline{K}}-\underline{\underline{K}}'-\underline{\underline{\underline{Aut}}}^W_{obs}$ of all observable automata is
realizing. We only have to show that each observable automa-
ton is minimal in $\underline{\underline{K}}-\underline{\underline{K}}'-\underline{\underline{\underline{Aut}}}^W$ such that $\underline{\underline{K}}-\underline{\underline{K}}'-\underline{\underline{\underline{Aut}}}^W_{obs}$ is minimal and hence reduced by 3.6,3.
Given an observable automaton A' and an arbitrary automa-
ton A such that $E(A) \subseteq E(A')$, i.e. there is a $\underline{\underline{K}}'$-morphism

$g:E(A) \to E(A')$ satisfying $m(A')\circ g = m(A)$, the morphism
$e(A'):S' \to E(A')$ is an isomorphism because A' is observable
and hence $M(A')\in\mathfrak{M}$. Thus we define $f:= e(A')^{-1}\circ g\circ e(A):S \to S'$
which is a weak morphism $f:A \to A'$ because we have
$M(A')\circ f = M(A)$ in diagram (7.8.2).

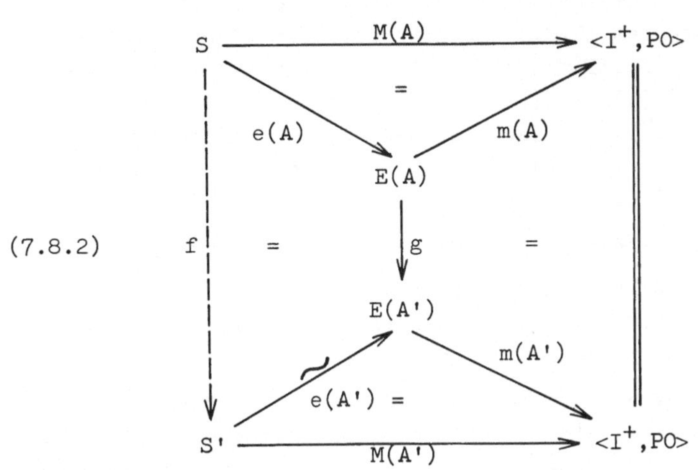

Moreover we have $f\in\mathfrak{C}$ in the case $g = id_{E(A)}$ such that A' is
minimal in $\underline{K}\text{-}\underline{K}'\text{-}\underline{\underline{Aut}}^w$.

Now we are going to prove the additional assertions:

1. is proved in 7.4.

2. One part has been verified above and the other part is
shown in 3.6,2.

3. Since observable systems are minimal it follows that
equivalent observable systems are isomorphic in $\underline{K}\text{-}\underline{K}'\text{-}\underline{\underline{Aut}}^w$ by
3.2 which implies that the state objects are isomorphic.

4. Let $f:A \to A'$ be a reduction with observable A' and c
a coretraction of $f:S \to S'$ in $\underline{K}$.
Since f is an automata morphism we have by (7.6.1):

$$f\circ d = d'\circ(f\otimes I) = d'\circ(f\otimes I)\circ(c\otimes I)\circ(f\otimes I) = f\circ d\circ(c\otimes I)\circ(f\otimes I) \quad .$$

If f can be replaced by e(A) we have exactly (7.8.1). In
fact we have M(A')∘f = M(A) by (7.6.4) which is an $\mathfrak{E}$-$\mathfrak{M}$-fac-
torization of M(A) because we have M(A')∈$\mathfrak{M}$ and f∈$\mathfrak{E}$ by
assumption. On the other hand we have the canonical $\mathfrak{E}$-$\mathfrak{M}$-fac-
torization m(A)∘e(A) = M(A) such that f coincides with
e(A) up to isomorphism. Combining the coretraction c with
this isomorphism (7.8.1) is also satisfied for e(A) .
Vice versa, assuming that e(A) satisfies (7.8.1) we have
$M(A_c)$∘e(A) = M(A) by (7.4.4) which is equivalent to
$(1_c)_n$∘$(e(A)⊗I^n) = 1_n$ for all n∈$\mathbb{N}$ (cf. (7.6.2),(7.6.4)) and
hence implies 1_c∘(e(A)⊗I) = 1 . On the other hand using
(7.4.1) and (7.8.1) we get

$$d_c∘(e(A)⊗I) = e(A)∘d∘(c⊗I)∘(e(A)⊗I) = e(A)∘d ,$$

which shows that E(A) is compatible with d and d_c , and
hence an automata morphism $e(A):A → A_c$ which is a reduction from A to the observable automaton A_c (cf. 7.4).

5. Let A be observable and f:A → A' a reduction. By
(7.6.2) we have M(A')∘f = M(A)∈$\mathfrak{M}$ which implies f∈$\mathfrak{E}$∩$\mathfrak{M}$ by
4.7,3. Hence f is a $\underline{K}$'-isomorphism by 4.7,4 and thus an
isomorphism in $\underline{K}$-$\underline{K}$'-$\underline{Aut}$ showing that A is reduced.
Now let A be a reduced automaton satisfying (7.8.1) then
$e(A):A → A_c$ is a reduction as shown above and hence an isomorphism because A is reduced. Thus $A ≅ A_c$ is observable.
Vice versa each observable automaton satisfies (7.8.1) because e(A) is an isomorphism with inverse morphism c .

6. is an application of 3.7 to the subsystematic $\underline{K}$-$\underline{K}$'-$\underline{Aut}^W_{obs}$
which has been shown to be minimal and realizing.   ■

<u>7.9 Strong Minimality</u>: Since equivalent observable automata
are not necessarily isomorphic (cf. 7.8,3) it is natural to
ask for a subclass having this property. Such a class is defined by strong observable automata, meaning that not only
the states but also the subsets of states are pairwise inequivalent in the case of nondeterministic automata. In
other words the function u∘℘'M(A):℘'S → <I^+,℘'O> has to be

injective where u is the union (cf.6.5).

<u>Definition</u>: An observable automaton A is called <u>strong</u>
<u>observable</u> if the morphism

$$PS \xrightarrow{\;PM(A)\;} P<I^+,PO> \xrightarrow{\;u\;} <I^+,PO>$$

belongs to $\mathfrak{M}$ (cf. (6.5.1),(6.6.2)).

<u>Remark</u>: In most examples the unit morphisms $i_S:S \to PS$ of
the adjunction $J \dashv P$ (cf. 6.1) belong to the class $\mathfrak{M}$. In
this case $u \cdot PM(A) \in \mathfrak{M}$ implies that A is also observable
because we have $M(A) = u \cdot PM(A) \cdot i_S$ (cf. 8.3,2).

<u>Example</u>: We now consider an example of a strong observable
automaton in the nondeterministic case.
Given the automaton $A_7 = (\{x\},\{y_1,y_2,y_3\},S_7=\{1,2,3\},d_7,l_7)$
by (7.9.1)

(7.9.1)

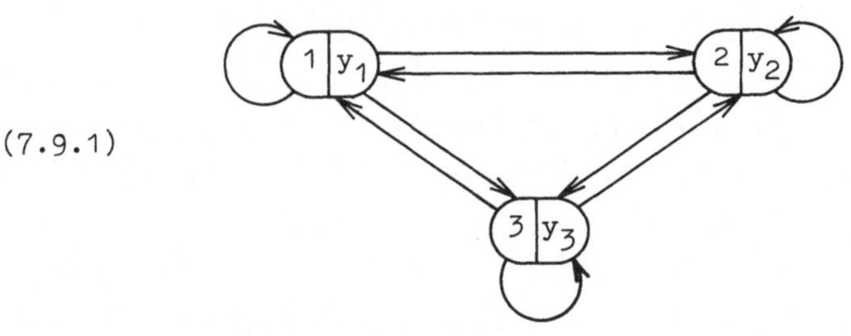

calculation of the machine function $M(A_7)$ yields for
$i = 1,2,3$:

$$M(A_7)(i)(x^n) = 1_7^+(i,x^n) = \begin{cases} \{y_i\} & \text{for n=1} \\ \{y_1,y_2,y_3\} & \text{otherwise,} \end{cases}$$

and the corresponding input-output behavior of subsets S
of S_7 is defined by:

$$u \cdot P'M(A_7)(S)(x^n) = \bigcup_{i \in S} M(A_7)(i)(x^n) = \begin{cases} \{y_i / \; i \in S\} & \text{for n=1} \\ \{y_1,y_2,y_3\} & \text{otherwise.} \end{cases}$$

Regarding the output for n = 1 we see that all subsets of

states are pairwise inequivalent and hence A_7 is strong observable. Note that the nondeterministic automata A_2 to A_6 , given in 2.7 and 6.4 , and all other equivalent automata do not satisfy this condition because their behavior is closed under union.

Theorem: Each strong observable automaton $\tilde{A}$ in $(\underline{K},\otimes)$ is minimal in the systematic $\underline{K}$-$\underline{K}'$-$\underline{\underline{Aut}}$.
In other words there is a unique automata morphism $f:A \to \tilde{A}$ for each automaton A with $E(A) \subseteq E(\tilde{A})$ which is a reduction in the case $E(A) = E(\tilde{A})$. Especially we have that equivalent strong observable automata are isomorphic.

Proof: Let $E(A) \subseteq E(\tilde{A})$ and $\tilde{A}$ strong observable. Thus $\tilde{A}$ is observable and according to 7.8,2 there is exactly one weak morphism $f:A \to \tilde{A}$ which belongs to $\mathfrak{S}$ in the case $E(A)=E(\tilde{A})$. Thus we have

$$(7.9.2) \qquad\qquad M(\tilde{A}) \circ f = M(A)$$

which implies $\tilde{1} \circ (f \otimes I) = 1$ such that it remains to show

$$(7.9.3) \qquad\qquad \tilde{d} \circ (f \otimes I) = f \circ d \ .$$

By (6.2.1) there are unique $\underline{K}'$-morphisms d' resp. $\tilde{d}'$ satisfying $v_S \circ d' = d$ and $v_{\tilde{S}} \circ \tilde{d}' = \tilde{d}$ such that (7.9.3) is equivalent to

$$(7.9.4) \qquad\qquad \tilde{d}' \circ (f \otimes I) = Pf \circ d' \qquad\quad (cf.(6.6.2)).$$

Since $\tilde{A}$ strong observable implies that $u \circ PM(\tilde{A})$ is a monomorphism, (7.9.4) can be obtained from the following diagram (7.9.5) which commutes by (7.9.2) and (6.6.1).

S⊗I
d'
PS
Pf
PS̃
(7.9.2)
f⊗I
M(A)⊗I
PM(A)
PM(Ã)
S̃⊗I
M(Ã)⊗I
<I⁺,PO>⊗I
(6.6.1)
P<I⁺,PO>
d̃'
(6.6.1)
L
u
PS̃
PM(Ã)
P<I⁺,PO>
u
<I⁺,PO>

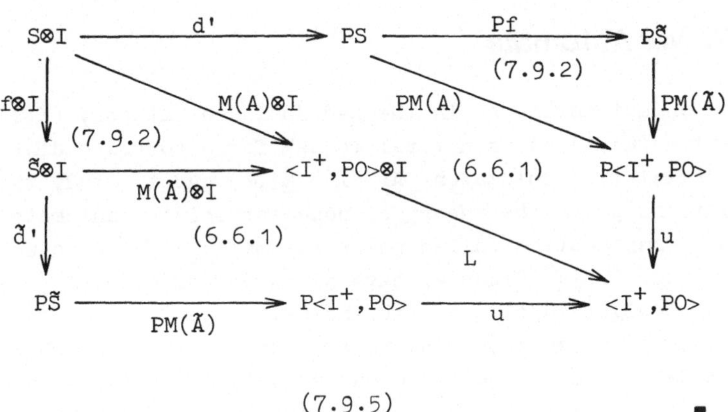

(7.9.5) ∎

8. Power Automata

Given an automaton A in the pseudoclosed category $(\underline{K},\otimes)$
relative $(\underline{K}',\otimes)$ it is natural to ask for a corresponding
"deterministic" automaton A' in $(\underline{K}',\otimes)$ and to study their
relationship. In the theory of nondeterministic automata
such a construction called power automaton $\tilde{P}A$ of A is well-
known (cf. [75]). Since we have already a functor $P:\underline{K} \to \underline{K}'$
which is right adjoint to the inclusion functor from $\underline{K}'$ to $\underline{K}$
it is easy to construct the power automaton $\tilde{P}A$ for arbitrary
automata A in a pseudoclosed category $(\underline{K},\otimes)$. $\tilde{P}$ turns out to
be a functor from the category of automata in $(\underline{K},\otimes)$ to those
in $(\underline{K}',\otimes)$ which is right adjoint to the inclusion again.
The behavior $E(\tilde{P}A)$ of $\tilde{P}A$ can be shown to be the union
closure of $E(A)$ and to have other interesting closure properties (cf. 8.3). As a corollary we get the result that an
automaton A is strong observable iff the power automaton
$\tilde{P}A$ is observable (cf. 8.4). Finally we will make some remarks concerning the construction of the kernel automaton
$\check{K}A$ which is the smallest subautomaton of $\tilde{P}A$ including A
(cf. 8.5).

<u>8.1 General Assumptions</u>: According to 6.1 $(\underline{K},\otimes)$ will be a
pseudoclosed category relative $(\underline{K}',\otimes)$ where $(\underline{K}',\otimes)$ is assumed to be closed, to have countable coproducts and an
$\mathfrak{C}$-$\mathfrak{M}$-factorization. As before $P:\underline{K} \to \underline{K}'$ will be the right
adjoint functor to the inclusion functor $J:\underline{K}' \to \underline{K}$.
Since the power automaton $\tilde{P}A$ will have the output object
PO the corresponding category of automata in $(\underline{K}',\otimes)$ with
input object I and output object PO will be denoted by
$\underline{K}'$-<u>Aut</u>(I,PO). On the other hand $\underline{K}$-<u>Aut</u> will denote the category of automata in $(\underline{K},\otimes)$ with fixed I and O as before,
but morphisms are not restricted to be $\underline{K}'$-morphisms any
longer (cf. 7.1).

Finally $\underline{K}'\text{-}\underline{Aut}(I,PO)$ can be regarded as a subcategory of $\underline{K}\text{-}\underline{Aut}$ because each $A'=(S,d,l')$ in $\underline{K}'\text{-}\underline{Aut}(I,PO)$ with $l':S\otimes I \to PO$ can be represented as an automaton $A=(S,d,l)$ in $\underline{K}\text{-}\underline{Aut}$ with $l:=v\circ l':S\otimes I \to O$ in $\underline{K}$ (cf. (6.2.1)). Vice versa each automaton $A=(S,d,l)$ in $\underline{K}\text{-}\underline{Aut}$ with $d\in\underline{K}'$ can be regarded as an automaton $A'=(S,d,l')$ in $\underline{K}'\text{-}\underline{Aut}(I,PO)$. Let us remark that in both cases the constructions for the machine morphism and the behavior coincide, i.e. $M(A)=M(A')$ and $E(A)=E(A')$.

<u>8.2 Definition</u> (P o w e r   A u t o m a t a) : Given an automaton $A=(S,d,l)$ in $\underline{K}\text{-}\underline{Aut}$ the automaton $\check{P}A=(PS,\tilde{d},\tilde{l})$ in $\underline{K}'\text{-}\underline{Aut}(I,PO)$ is called <u>power automaton</u> of A where $\tilde{d}$ and $\tilde{l}$ are defined by the universal properties of v_S and v in (8.2.1) (cf. (6.2.1)):

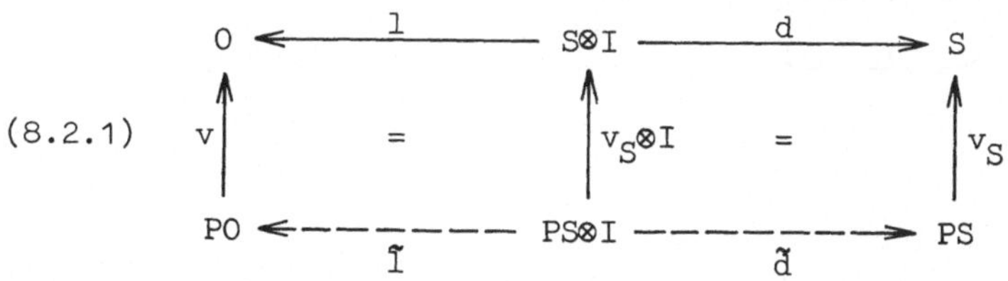

$$(8.2.1)$$

This construction implies that $v_S:PS \to S$ is a morphism $v_S:\check{P}A \to A$ in $\underline{K}\text{-}\underline{Aut}$.

<u>Interpretation</u>: Given a nondeterministic automaton $A=(S,d,l)$ the state object PS of $\check{P}A$ is the power set $P'S$ of S without the empty subset and for each $M\subseteq S$ and $x\in I$ we have by definition of v_S and v in 6.2 example 1:

$$\tilde{d}(M,x) = \bigcup_{s\in M} d(s,x) \quad \text{and} \quad \tilde{l}(M,x) = \bigcup_{s\in M} l(s,x) \ .$$

Replacing P' by P the same interpretation is true for relational and relational topological automata whereas in the case of stochastic automata PS is the set of all probabil-

ity distributions on S and for each $p \in PS$, $x \in I$, $s' \in S$, $y \in O$
we have
$$\tilde{d}(p,x)(s') = \sum_{s \in S} p(s) \cdot d(s,x)(s')$$
and
$$\tilde{l}(p,x)(s') = \sum_{s \in S} p(s) \cdot l(s,x)(s') \ .$$

<u>Example</u>: Given the nondeterministic automaton A_7 in (7.9.1)
the power automaton $\tilde{P}A_7$ is given by the following graph:

(8.2.2)

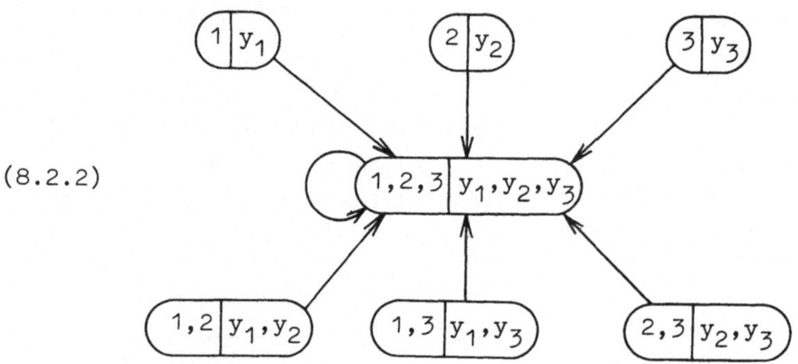

<u>8.3 Theorem</u> (P o w e r   A u t o m a t a) :
Let $\tilde{P}A = (PS, \tilde{d}, \tilde{l})$ be the power automaton of $A = (S,d,l)$ and
$v_S : \tilde{P}A \to A$ given by (8.2.1) then we have:

1. For each automata morphism $f : A' \to A$ in <u>K</u>-<u>Aut</u> with A' in
<u>K</u>'-<u>Aut</u>(I,PO) there is a unique morphism $f' : A' \to \tilde{P}A$ in
<u>K</u>'-<u>Aut</u>(I,PO) such that diagram (8.3.1) commutes:

(8.3.1)

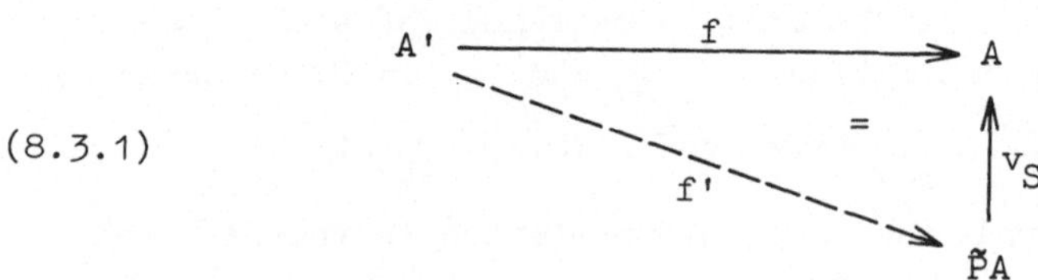

P̃ becomes a functor P̃:$\underline{K}$-$\underline{Aut}$ → $\underline{K}$'-$\underline{Aut}$(I,PO) which is right adjoint to the inclusion J:$\underline{K}$'-$\underline{Aut}$(I,PO) → $\underline{K}$-$\underline{Aut}$ (cf. 8.1).

2. We have the following correspondence between the machine morphisms M(P̃A) and M(A) :

  (i)

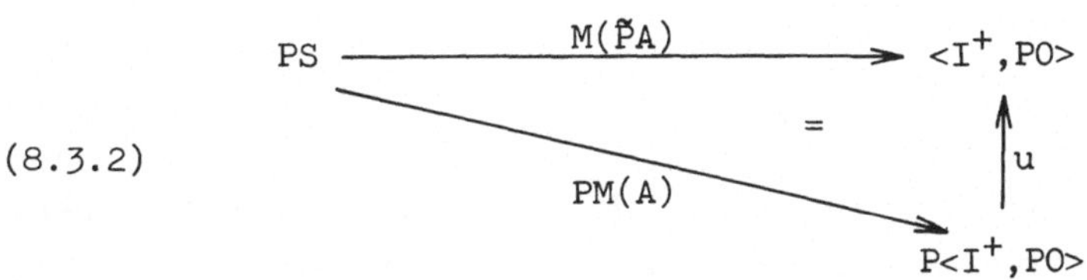

(8.3.2)

where u is the union defined in 6.5 and PM(A) is the application of P to M(A) (cf. (6.6.2)).

  (ii)

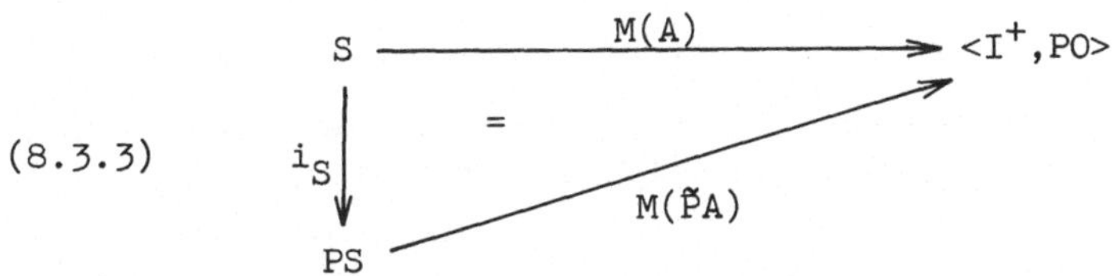

(8.3.3)

where i_S is the unit of the adjunction J⊣P defined by $v_S \circ i_S = id_S$ (cf. (6.2.1)).

3. We have the following closure properties:

    (i) E(P̃A) is the u-closure of E(A) meaning that there is a morphism e∈𝕰 such that (8.3.4) is a canonical 𝕰-𝔐-factorization of u∘Pm(A) where m(A) and similarly m(P̃A) are defined by (6.3.4).

   (ii) E(A) ⊆ E(P̃A) = E(P̃P̃A)

 (iii) For all automata A' in $\underline{K}$-$\underline{Aut}$, E(A) ⊆ E(A') implies E(P̃A) ⊆ E(P̃A') .

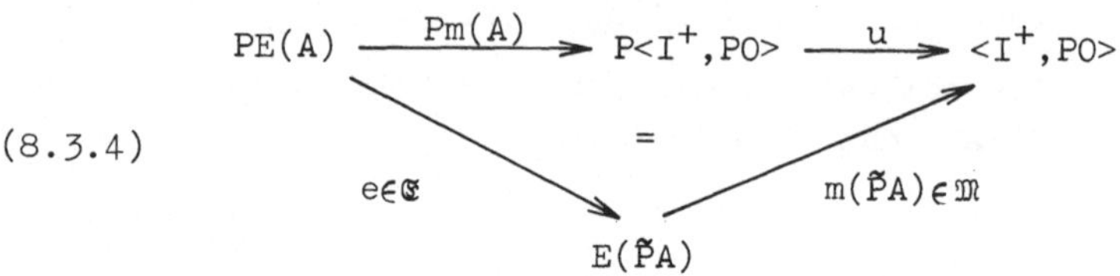

$$\text{(8.3.4)}$$

Diagram: $PE(A) \xrightarrow{\ Pm(A)\ } P\langle I^+, PO\rangle \xrightarrow{\ u\ } \langle I^+, PO\rangle$, with $e \in \mathfrak{C}$ and $m(\check{P}A) \in \mathfrak{M}$ meeting at $E(\check{P}A)$ and $=$.

<u>Remark</u>: For the assertions 3.(i) and 3.(iii) we assume that the class $\mathfrak{C}$ is preserved by P , i.e. $Pe \in \mathfrak{C}$ for all $e \in \mathfrak{C}$. This condition is satisfied in all our examples.

<u>Interpretation</u>: The power automaton $\check{P}A$ of A is characterized up to isomorphism by the universal property (8.3.1) which is based on (6.2.1). According to the interpretation of the union in 6.5 the machine morphism $M(\check{P}A)$ of $\check{P}A$, which is equal to $u \circ PM(A)$ by (8.3.2), satisfies

$$M(\check{P}A)(S') = \bigcup_{s \in S'} M(A)(s) \qquad \text{for all } S' \subseteq S$$

in the case of nondeterministic automata for example. Hence the input-output behavior of a subset $S' \in P'S$ is the union of the behaviors $M(A)(s)$ of all the states $s \in S'$. Vice versa we have $M(\check{P}A)(\{s\}) = M(A)(s)$ for a single element subset $S' = \{s\}$ by (8.3.3) because $i_S : S \to P'S$ is defined by $i_S(s) = \{s\}$. Finally (8.3.4) asserts that the behavior $E(\check{P}A)$ of $\check{P}A$ is the closure of $E(A)$ under union, e.g. in (8.2.2) we have:

$$E(\check{P}A_7) = \{f_1, f_2, f_3, f_1 \cup f_2, f_2 \cup f_3, f_1 \cup f_3, f_1 \cup f_2 \cup f_3\}$$

for $E(A_7) = \{f_i := M(A_7)(i)/\ i = 1,2,3\}$ (cf. example 7.9). Conditions 3.(ii) and (iii) are the usual closure properties.

<u>Proof of the theorem</u>: 1. Given A' in $\underline{K}'\text{-}\underline{Aut}(I, PO)$ and $f : A' \to A$ in $\underline{K}\text{-}\underline{Aut}$ there is a unique $\underline{K}'$-morphism $f' : S' \to PS$ satisfying $v_S \circ f' = f$ by (6.2.1). In order to show (8.3.1) it remains to verify that f' is already an automata mor-

phism $f':A' \to \tilde{P}A$ which follows from diagram (8.3.5) below.

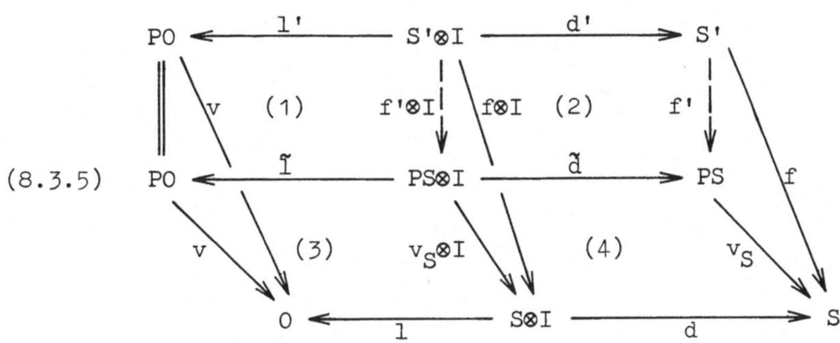

Subdiagrams (3) and (4) in (8.3.5) are commutative by
(8.2.1) and the triangles by definition of f' showing that
(1) and (2) are equalized by v and v_S respectively and
hence commutative by the uniqueness of the supplement in
(6.2.1).

2. (i) Using the universal properties of $v \circ ev$ in (6.2.2)
condition (8.3.2) is equivalent to the commutativity of diagram (8.3.6).

   (ii) Using (8.3.6) below and $v_S \circ i_S = id_S$ we have

$$(v \circ ev) \circ (M(\tilde{P}A) \otimes I^+) \circ (i_S \otimes I^+) = v \circ ev \circ (M(A) \otimes I^+) \circ (v_S \otimes I^+) \circ (i_S \otimes I^+)$$
$$= v \circ ev \circ (M(A) \otimes I^+)$$

which implies $M(\tilde{P}A) \circ i_S = M(A)$ by the uniqueness of $M(A)$
defined in (6.3.3).

3. (i) Define e to be the unique diagonal morphism
(cf. 4.7,1) in the diagram (8.3.7) which commutes by (8.3.2)
and (6.3.4).
$e(\tilde{P}A) \in \mathfrak{E}$ in (8.3.7) implies $e \in \mathfrak{E}$ such that $m(\tilde{P}A) \circ e$ is the
canonical $\mathfrak{E}$–$\mathfrak{M}$-factorization of $u \circ Pm(A)$.

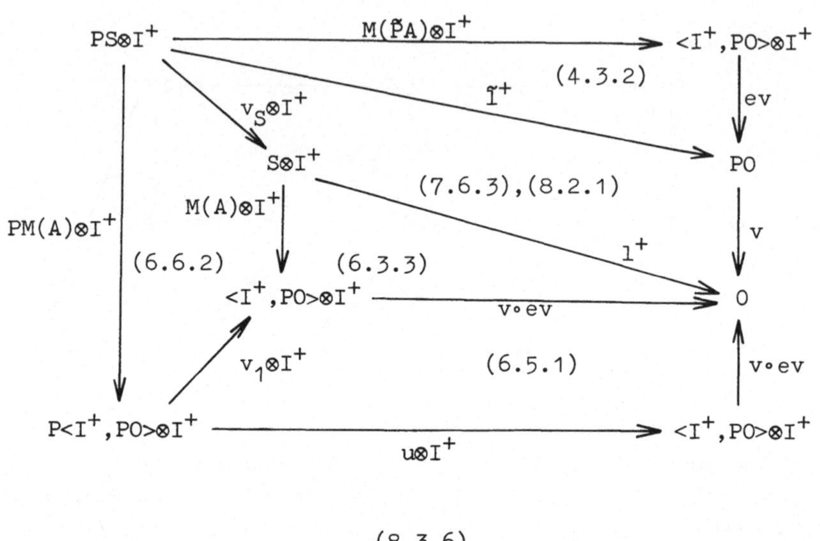

(8.3.6)

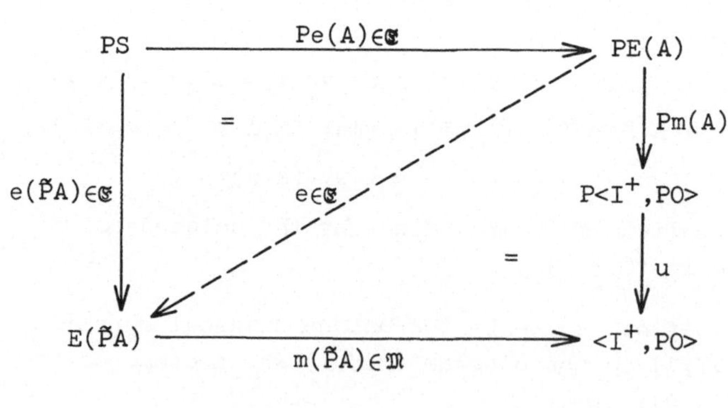

(8.3.7)

147

(ii) Since we have canonical $\mathfrak{E}$-$\mathfrak{M}$-factorizations $m(A) \circ e(A) =$
$= M(A)$ and $m(\check{P}A) \circ e(\check{P}A) = M(\check{P}A)$ we can apply the diagonal
lemma 4.7,1 to (8.3.3) showing that there is a morphism
$d : E(A) \to E(\check{P}A)$ such that $m(\check{P}A) \circ d = m(A)$ and hence
$E(A) \subseteq E(\check{P}A)$ by definition of the partial ordering in the
behavior category $\underline{B}$ (cf. 7.6).
Replacing A by $\check{P}A$ we get $E(\check{P}A) \subseteq E(\check{P}\check{P}A)$. Vice versa
$E(\check{P}\check{P}A) \subseteq E(\check{P}A)$ follows from the fact that $\check{P}v_S : \check{P}\check{P}A \to \check{P}A$ is
an automata morphism and hence $E\check{P}v_S : E(\check{P}\check{P}A) \to E(\check{P}A)$ is a
behavior morphism because $\check{P}$ and E are functors (cf.
8.3,1 and 5.2,2).

(iii) $E(A) \subseteq E(A')$ means that there is a $\underline{K}'$-morphism
$m : E(A) \to E(A')$ satisfying $m(A') \circ m = m(A)$. Thus we have

$$(8.3.8) \qquad u \circ Pm(A) = u \circ Pm(A') \circ Pm.$$

Since $E(\check{P}A)$ is the image of $M(\check{P}A) = u \circ PM(A) = u \circ Pm(A) \circ Pe(A)$
and $Pe(A) \in \mathfrak{E}$ it follows that $E(\check{P}A)$ is the image of
$u \circ Pm(A)$ (cf. 4.6). Similarly $E(\check{P}A')$ is the image of
$u \circ Pm(A')$ which implies $E(\check{P}A) \subseteq E(\check{P}A')$ using the diagonal
lemma 4.7,1 in (8.3.8). ∎

8.4 Corollary (O b s e r v a b i l i t y) : An automaton
A was called strong observable in 7.9 if the morphism
$u \circ PM(A)$ belongs to the class $\mathfrak{M}$. Using $u \circ PM(A) = M(\check{P}A)$ (cf.
(8.3.2)) we have the following corollary:

An automaton A is strong observable iff the power automaton $\check{P}A$ is observable, i.e. $M(\check{P}A) \in \mathfrak{M}$.

Remark: Since $\check{P}A$ is an automaton in the closed category
$(\underline{K}', \otimes)$ we have up to isomorphism a unique observable automaton A' in $(\underline{K}', \otimes)$ which is equivalent to $\check{P}A$. Unfortunately there is no automaton A'' in $(\underline{K}, \otimes)$ such that
$\check{P}A'' = A'$ in general. Otherwise this would be a method to
construct for each automaton A in $(\underline{K}, \otimes)$ an equivalent
strong observable automaton A'' which is impossible (cf. 7.9).
On the other hand the power automaton is used for algorithms

for the minimization of nondeterministic automata in [70].
The minimization of the power automaton in the case of initial automata corresponds to the minimal realization construction of "nondeterministic" machines in [6] leading to
the minimal deterministic machine which is equivalent to the
given nondeterministic machine. Note that nondeterministic
machines correspond to automata in pseudoclosed categories
in our sense (cf. remark in 6.2).

<u>8.5 Remark</u> (K e r n e l   A u t o m a t a) : According to
(8.3.3) the power automaton $\tilde{P}A$ "simulates" the nondeterministic automaton A , but the smallest deterministic automaton simulating A is the subautomaton $\check{K}A$ of $\tilde{P}A$, generated by the states of A, which is called kernel automaton
$\check{K}A$ of A. The kernel automaton $\check{K}A_7$ of A_7 in (7.9.1) is
given by the following subautomaton of $\tilde{P}A_7$ in (8.2.2) for
example.

(8.5.1)

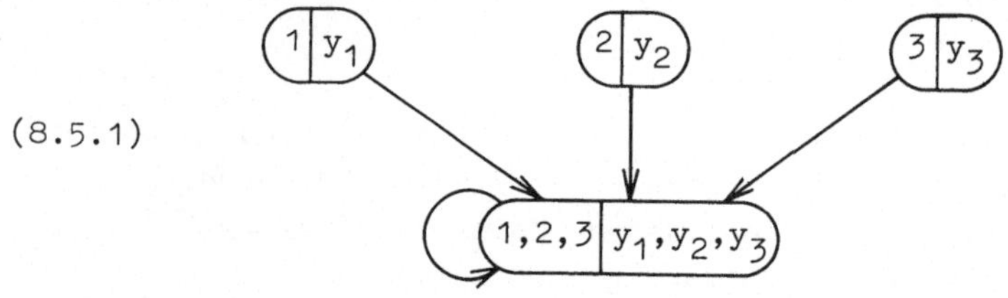

We only want to give the construction of the kernel automaton $\check{K}A$ in the case of a nondeterministic automaton
$A = (S,d,1)$. $\check{K}A$ is given by $\check{K}A = (\bar{S},\bar{d},I)$ with

$$\bar{S} := \{\tilde{d}*(\{s\},w)\in P'S/\ s\in S, w\in I*\} \subseteq P'S$$

and $\bar{d}$ resp. I are the restrictions of $\tilde{d}$ and $\tilde{1}$ to $\bar{S}xI$
respectively where $\tilde{P}A = (P'S,\tilde{d},\tilde{1})$ is the power automaton of
A and $\tilde{d}*:P'SxI* \to P'S$ is defined in 1.2.

The kernel automaton $\check{K}A$ of A has the following properties:

1. $\check{K}A$ is the smallest subautomaton of $\check{P}A$ containing all
the states $s \in S$ of A , or more precisely all subsets
$\{s\} \in P'S$ with $s \in S$.

2. The behavior $E(\check{K}A)$ of the kernel automaton is the
smallest behavior of a deterministic automaton which in-
cludes the behavior $E(A)$ of A , i.e. $E(A) \subseteq E(\check{K}A) \subseteq E(\check{P}A)$
and for each deterministic automaton A' in $\underline{Aut}(I,P'O)$ satisfying $E(A) \subseteq E(A')$ we have already $E(\check{K}A) \subseteq E(A')$.

3. $E(\check{K}A)$ is the closure of $E(A)$ under left shift L_w with
strings $w \in I^*$, i.e.

$$E(\check{K}A) = \{ f \circ L_w : I^+ \to P'O/ \ f \in E(A), w \in I^* \}$$

where $L_w : I^+ \to I^+$ is defined by $L_w(v) = wv$ for all $v \in I^+$.

A generalization of these constructions and results for
automata in pseudoclosed categories together with the corresponding proofs is given in [57].

9. Initial Automata

In this chapter we will extend our theory to automata with
fixed initial state, called initial automata. The initial
state will be given by a $\underline{K}$-morphism $a:U \to S$ where U is
the unit object in the monoidal category $(\underline{K},\otimes,U)$ and S the
state object of an automaton. If A is a deterministic type
automaton and $U = \{1\}$ the initial state is for example defined by $a(1)\in S$. The behavior of an initial automaton is
defined as being the input-output morphism of the initial
state.
Most of our constructions concerning reduction and minimization can be extended to initial automata. For the case of
automata in closed categories we get the same strong results
as given in chapters 4 and 5, provided that we have reachable automata meaning that each state is reachable from the
initial state. Otherwise we have to start with a reachability construction leading to an equivalent reachable automaton
(cf. 9.7).
The construction of an equivalent reachable and observable
automaton in a pseudoclosed category, however, does not lead
to an automaton with a minimal number of states in the initial case.
For the realization of input-output morphisms we give a free
and an observable realization construction which, of course,
is minimal only in the case of initial automata in closed
categories. Moreover we are able to characterize those input-output morphisms which are realized by initial automata
with finite state object.

Since several constructions in this chapter together with
their automata theoretic interpretation are similar to those
in the preceding chapters we will give a more comprehensive
presentation of the theory in this section and some proofs
will only be sketched.

For examples we refer to 4.9 and 6.8 which can of course be
extended to the initial case.

9.1 <u>General Assumptions</u>: Let $(\underline{K}',\otimes)$ be a closed category
with countable coproducts and $\mathfrak{E}$-$\mathfrak{M}$-factorization and $(\underline{K},\otimes)$ a
pseudoclosed category relative $(\underline{K}',\otimes)$ in the sense of 6.1.
For the construction of reachable automata we will assume
also having an $\tilde{\mathfrak{E}}$-$\mathfrak{M}$-factorization in $(\underline{K},\otimes)$ in the sense of
(4.6.1) with the same class $\mathfrak{M}$ as above and a class of "sur-
jective" morphisms $\tilde{\mathfrak{E}}$ in $\underline{K}$ such that $\tilde{\mathfrak{E}}$ restricted to $\underline{K}'$ is
equal to $\mathfrak{E}$. But $\tilde{\mathfrak{E}}$ is not assumed to consist of epimorphisms
in $\underline{K}$ taking for example surjective nondeterministic functions and relations in the categories <u>ND</u> and <u>Rel</u> respectively.
Finally $\underline{\underline{K}}\text{-}\underline{\underline{Aut}}^{\cdot} = (\underline{K}\text{-}\underline{Aut}^{\cdot},\underline{B}^{\cdot},E^{\cdot})$ will denote the systematic
of initial automata in $(\underline{K},\otimes)$ which will be introduced in
9.2, and similar to 7.1 $\underline{K}\text{-}\underline{K}'\text{-}\underline{Aut}^{\cdot}$ will be the category of
initial automata with morphisms in $\underline{K}'$.
We do not need all these assumptions for all constructions
in this chapter but it is obvious which of them are necessary in each case.

9.2 <u>Definition</u> (S y s t e m a t i c   o f   I n i t i a l
A u t o m a t a) : An <u>initial automaton</u> $A = (S,d,l,a)$ in
$(\underline{K},\otimes)$ consists of a $\underline{K}$-object S and $\underline{K}$-morphisms $d:S\otimes I \to S$,
$l:S\otimes I \to O$ and $a:U \to S$ where I and O are fixed input
and output objects and U is the unit object in $(\underline{K},\otimes)$.
Given initial automata A and A' a $\underline{K}$-morphism $f:S \to S'$
is called <u>automata morphism</u> $f:A \to A'$ if (9.2.1) is commutative.
Initial automata in $(\underline{K},\otimes)$ together with these automata mor-
phisms constitute the <u>category $\underline{K}\text{-}\underline{Aut}^{\cdot}$</u> of <u>initial automata</u>
in $(\underline{K},\otimes)$.
The <u>behavior</u> $E^{\cdot}(A)$ of an initial automaton A is given by

$$(9.2.2) \qquad E^{\cdot}(A) = (I^{+} = U\otimes I^{+} \xrightarrow{\ a\otimes I^{+}\ } S\otimes I^{+} \xrightarrow{\ l^{+}\ } O)$$

where $l^{+}:S\otimes I^{+} \to O$ is defined in (6.3.2).

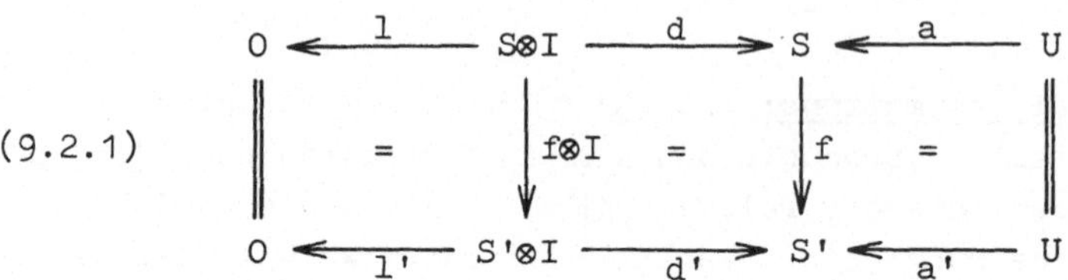

$$(9.2.1)$$

Since each automata morphism $f:A \to A'$ in $\underline{K}$-$\underline{Aut}^{\cdot}$ implies $E^{\cdot}(A) = E^{\cdot}(A')$ by

$$(9.2.3)$$

$E^{\cdot}$ becomes a <u>behavior functor</u> $E^{\cdot}:\underline{K}$-$\underline{Aut}^{\cdot} \to \underline{B}^{\cdot}$ where $\underline{B}^{\cdot}$ is the discrete <u>behavior category</u> having as objects all $\underline{K}$-morphisms $b:I^{+} \to 0$.

Hence $\underline{\underline{K}}$-$\underline{\underline{Aut}}^{\cdot} = (\underline{K}$-$\underline{Aut}^{\cdot},\underline{B}^{\cdot},E^{\cdot})$ is a systematic in the sense of 3.1, called <u>systematic of initial automata</u> in $(\underline{K},\otimes)$. Restricting the morphisms f in (9.2.1) to be $\underline{K}'$-morphisms the corresponding category and systematic will be denoted by $\underline{\underline{K}}$-$\underline{\underline{K}}'$-$\underline{\underline{Aut}}^{\cdot} = (\underline{K}$-$\underline{K}'$-$\underline{Aut}^{\cdot},\underline{B}^{\cdot},E^{\cdot})$. In the case $\underline{K} = \underline{K}'$ both notions coincide leading to the concept of <u>initial automata in closed categories</u>.

<u>Remark</u>: The equality $I^{+} = U \otimes I^{+}$ in (9.2.2) in fact is only a natural isomorphism $1_{I^{+}}:U \otimes I^{+} \overset{\sim}{\to} I^{+}$ but due to our convention in 1.9 we will not note these isomorphisms explicitly.

<u>Examples and Interpretation</u>: All examples of automata given in the closed and pseudoclosed cases (cf. 4.9 and 6.8)

153

can be extended to initial automata. In the deterministic
cases with $U = \{1\}$ or $U = R$ for $\underline{Mod}_R$ $\quad a:U \to S$ assigns
one initial state $a(1) \in S$ whereas in the nondeterministic
cases we get a subset $a(1) \subseteq S$ of initial states. However,
for partial functions and relations $a(1)$ may be defined or
empty. The behavior $E^{\cdot}(A)$ in each case is the input-output
behavior $1^+(a(1),-):I^+ \to 0$ induced by the initial state or
states.

<u>9.3 Free Realization and Free Monoid</u>: Our aim is to show
that each behavior $b:I^+ \to 0$ in $\underline{B}^{\cdot}$ can be realized by an
initial automaton. A simple method to do that is the free
realization $F(b)$ which has as state object the <u>free monoid</u>
I^* (cf. 5.6 example 1) defined by the coproduct

$$(9.3.1) \qquad I^* = \coprod_{n \in \mathbb{N}_o} I^n \qquad \text{with} \quad I^o := U \ , \quad \mathbb{N}_o := \mathbb{N} \cup \{0\}$$

and coproduct injections $u_n:I^n \to I^*$ $(n \in \mathbb{N}_o)$.
The state transition morphism $u:I^* \otimes I \to I^*$ of the free realization is defined by

$$(9.3.2) \qquad u \circ (u_n \otimes I) = u_{n+1} \qquad \text{for all } n \in \mathbb{N}_o$$

using the fact that $u_n \otimes I:I^n \otimes I \to I^* \otimes I$ are coproduct injections (cf.4.2). Since we have

$$(9.3.3) \qquad I^* \otimes I = (\coprod_{n \in \mathbb{N}_o} I^n) \otimes I \cong \coprod_{n \in \mathbb{N}_o} I^{n+1} = I^+$$

the output morphism $1:I^* \otimes I \to 0$ of $F(b)$ is simply
$1 := (I^* \otimes I \cong I^+ \xrightarrow{b} 0)$ and the initial state is $u_o:U \to I^*$.

<u>Theorem</u> (F r e e   R e a l i z a t i o n) : Given a behav-
ior $b:I^+ \to 0$ in $\underline{B}^{\cdot}$ the automaton $F(b)$ defined by

$$(9.3.4) \qquad 0 \xleftarrow{\ b\ } I^+ \cong I^* \otimes I \xrightarrow{\ u\ } I^* \xleftarrow{\ u_o\ } U$$

realizes the behavior b, i.e. $E^{\cdot}F(b) := E^{\cdot}(F(b)) = b$ (cf. 12.5).
Moreover $F(b)$ is "reachable" in the sense defined below
and has the following universal properties showing that F
can be extended to a left adjoint functor of $E^{\cdot}:\underline{K}\text{-}\underline{Aut}^{\cdot} \to \underline{B}^{\cdot}$:
For each initial automaton $A = (S,d,1,a)$ in $\underline{K}\text{-}\underline{Aut}^{\cdot}$ with

154

$E^{\cdot}(A) = b$ there is a unique $\underline{K\text{-Aut}}^{\cdot}$-morphism $a^*:F(b) \to A$.
a^* is defined by the family

$$(9.3.5) \qquad a_o := a \ , \ a_{n+1} := d \circ (a_n \otimes I) \qquad (n \in \mathbb{N}_o)$$

which yields a unique $\underline{K}$-morphism $a^*:I^* \to S$ as induced morphism out of the coproduct I^* (cf. (4.2.0)) satisfying

$$(9.3.6) \qquad a^* \circ u_n = a_n \qquad \text{for all } n \in \mathbb{N}_o.$$

For the proof of the theorem we need (9.3.7) and (9.3.8) of the following lemma, which is straightforward to prove by induction whereas (9.3.9), which will be needed later, is a consequence of the theorem because $f \circ a^*$ and a'^* are both $\underline{K\text{-Aut}}^{\cdot}$-morphisms from $F(b)$ to A' and hence equal by uniqueness of such a morphism.

<u>Lemma</u>: Given an initial automaton $A = (S, d, l, a)$ we have (9.3.7) for a^* and u, defined by (9.3.6) resp. (9.3.2),

$$(9.3.7) \qquad
\begin{array}{ccc}
I^* \otimes I & \xrightarrow{\ \ u\ \ } & I^* \\
{\scriptstyle a^* \otimes I}\downarrow & = & \downarrow{\scriptstyle a^*} \\
S \otimes I & \xrightarrow{\ \ d\ \ } & S
\end{array}$$

meaning that a^* is a $\underline{K\text{-Medv}}$-automata morphism, and

$$(9.3.8) \qquad E^{\cdot}(A) = (I^+ \cong I^* \otimes I \xrightarrow{a^* \otimes I} S \otimes I \xrightarrow{l} 0)$$

which is an equivalent description of the behavior, and for each $f:A \to A'$ in $\underline{K\text{-Aut}}^{\cdot}$ we have

$$(9.3.9) \qquad f \circ a^* = a'^* : I^* \to S'$$

<u>Proof of the Theorem</u>: By uniqueness of a^* in (9.3.6) it follows that u_o^* is equal to id_{I^*} such that we have by (9.3.8):

$$E^{\cdot}F(b) = (I^+ \cong I^* \otimes I \xrightarrow{\mathrm{id}_{I^*} \otimes I} I^* \otimes I \cong I^+ \xrightarrow{b} 0) = b \ .$$

The $\underline{K}$-morphism $a^*:I^* \to S$ is an automata morphism
$a^*:F(b) \to A$ because we have $E^{\cdot}(A) = b$ in (9.3.10) below.

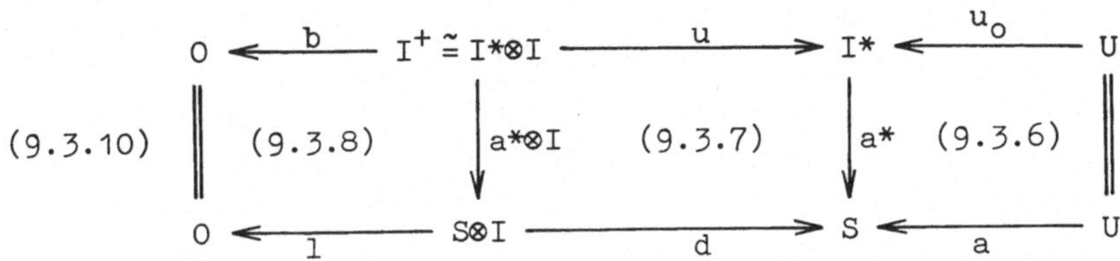

Uniqueness of a^* follows by induction from (9.3.10) using
(9.3.5) and (9.3.6). Finally $u_0^* = \mathrm{id}_{I^*} \in \mathfrak{C} \subseteq \tilde{\mathfrak{C}}$ implies that
$F(b)$ is reachable (cf. 9.4).   ∎

<u>9.4 Definitions</u>: Given a behavior $b:I^+ \to O$ in $\underline{B}^{\cdot}$ the auto-
maton $F(b)$ defined by (9.3.4) is called <u>free realization</u>
of b. For $A = (S,d,l,a)$ in $\underline{K\text{-}Aut}^{\cdot}$ $a^*:I^* \to S$ defined in
(9.3.6) is the <u>reachability morphism</u> of A and A is
called <u>reachable</u> if a^* belongs to the class $\tilde{\mathfrak{C}}$ in $\underline{K}$. Similar to the non-initial case, A is called <u>observable</u> if the
<u>machine morphism</u> M(A) of A constructed by (4.3.3) or
(6.3.3) belongs to the class $\mathfrak{M}$. A is <u>reduced</u> if each
<u>reduction</u> $f:A \to A'$, which is a $\underline{K\text{-}Aut}^{\cdot}$-morphism belonging
to the class $\mathfrak{C}$, is already an isomorphism.

<u>Interpretation</u>: According to the choice of $\tilde{\mathfrak{C}}$ in our exam-
ples (cf. 9.1) the reachability of A means that each state
of A can be reached by the initial state. The interpretation of observable and reduced automata corresponds to the
non-initial case.

<u>Comparison with the Non-Initial Case</u>: In order to compare
these notions for initial with those for non-initial automata we consider the forgetful functor

(9.4.1) $\qquad\qquad\qquad$ $V:\underline{K\text{-}Aut}^{\cdot} \to \underline{K\text{-}Aut}$

defined by $V(A^\cdot) = A$ for $A^\cdot = (S,d,l,a)$ in $\underline{K\text{-}Aut}^\cdot$ and $A = (S,d,l)$ in $\underline{K\text{-}Aut}$ and $V(f:A^\cdot \to A_1^\cdot) := f:A \to A_1$ (cf. (9.2.1)).

By definition we have that $A^\cdot$ is observable in $\underline{K\text{-}Aut}^\cdot$ iff $V(A^\cdot) = A$ has this property in $\underline{K\text{-}Aut}$ and each reduction $f:A^\cdot \to A_1^\cdot$ is also a reduction $f:A \to A_1$ in $\underline{K\text{-}Aut}$. Vice versa, for each reduction $f:V(A^\cdot) \to A_1$ in $\underline{K\text{-}Aut}$ there is a unique $A_1^\cdot$ in $\underline{K\text{-}Aut}^\cdot$ such that f becomes a reduction $f:A^\cdot \to A_1^\cdot$ in $\underline{K\text{-}Aut}^\cdot$ and hence diagram (9.4.2) commutes.

$$(9.4.2)\qquad
\begin{array}{ccc}
V(A^\cdot) & \xrightarrow{\ \ f\ \ } & A_1 \\[2pt]
 & \searrow\ {\scriptstyle V(f:A^\cdot \to A_1^\cdot)} & \Big\| \ {\scriptstyle =} \\[2pt]
 & & V(A_1^\cdot)
\end{array}$$

For $A_1 = (S_1,d_1,l_1)$ the initial automaton $A_1^\cdot$ is defined by

$$(9.4.3)\qquad A_1^\cdot = (S_1,d_1,l_1,a_1) \quad \text{with}\quad a_1 = f\circ a.$$

Moreover we have using (9.3.9) and $f \in \mathcal{C}$:

$(9.4.4)\qquad$ Reachability of $A^\cdot$ implies that of $A_1^\cdot$.

Finally (9.4.2) implies that $A^\cdot$ is reduced in $\underline{K\text{-}Aut}^\cdot$ iff $V(A^\cdot)$ is reduced in $\underline{K\text{-}Aut}$.

<u>Remark</u>: Note that the construction of $A_1^\cdot$ in (9.4.3) depends on $A^\cdot$ and f such that (9.4.2) is not the solution of the couniversal problem of V. But exactly (9.4.3) will be used to carry over the construction of reduced and observable automata in $\underline{K\text{-}K'\text{-}Aut}$ to the case of initial automata (cf. 9.8, 9.9).

Reachability is obviously a necessary condition for initial automata to have a minimal number of states, because the non- reachable states are superfluous with respect to the

behavior. Since the free realization is already reachable by
9.3 we will restrict ourselves to reachable automata in our
next theorem, but we will show in 9.7 that there is a couniversal construction to get the reachable part of an arbitrary initial automaton.

$\underline{9.5\ \text{Theorem}}$ (M i n i m a l   R e a l i z a t i o n
P r i n c i p l e   f o r   I n i t i a l   A u t o m a t a
i n   C l o s e d   C a t e g o r i e s) : Let $(\underline{K},\otimes) = (\underline{K}',\otimes)$
be a closed category then we have for the full subcategory
$\underline{K}\text{-}\underline{\text{Aut}}_{\otimes}^{\cdot}$ of all reachable automata in $\underline{K}\text{-}\underline{\text{Aut}}^{\cdot}$:
The subsystematic $\underline{\underline{K}}\text{-}\underline{\underline{\text{Aut}}}_{\otimes}^{\cdot}$ of $\underline{\underline{K}}\text{-}\underline{\underline{\text{Aut}}}^{\cdot}$ satisfies the Minimal
Realization Principle, i.e. for each behavior $b:I^{+} \to O$ in
$B^{\cdot}$ there is a minimal realization $M*(b)$ in $\underline{K}\text{-}\underline{\text{Aut}}_{\otimes}^{\cdot}$ with
$E^{\cdot}M*(b) = b$ such that for all A in $\underline{K}\text{-}\underline{\text{Aut}}_{\otimes}^{\cdot}$ with $E^{\cdot}(A) = b$
there is a unique reduction $f:A \to M*(b)$.

$\underline{\text{Remark}}$: The minimal realization $M*(b)$ turns out to be the
reduced automaton of the free realization $F(b)$. Note that
the main problems concerning reduction, minimization and behavior realization are solved by this theorem using 3.4 and
3.6. We do not state all these results explicitly because
they are exactly the same as in theorem 5.5 replacing non-
initial automata by reachable initial automata. Especially
we have that each reachable and observable automaton, such as
$M*(b)$, is minimal and hence has a minimal number of states
with respect to all equivalent initial automata.

$\underline{\text{Proof}}$: Given a behavior $b:I^{+} \to O$ the free realization $F(b)$
is reachable as shown in 9.3. Considering $VF(b)$ in $\underline{K}\text{-}\underline{\text{Aut}}$
(cf. (9.4.1)) we have a reduction $u:VF(b) \to RVF(b)$ in
$\underline{K}\text{-}\underline{\text{Aut}}$ by 5.5,2 which leads to a reduction $u:F(b) \to R^{\cdot}F(b)$
by (9.4.2) with reachable $R^{\cdot}F(b) := A_{1}^{\cdot}$ in (9.4.3) for
$A_{1} := RVF(b)$. Defining

$$(9.5.1) \qquad\qquad M*(b) := R^{\cdot}F(b)$$

we get a reachable realization of b because we have
$E^{\cdot}M*(b) = E^{\cdot}R^{\cdot}F(b) = E^{\cdot}F(b) = b$. Clearly $M*(b)$ is observable

and reduced by 9.4 and there is a unique reduction
$f:A \to M^*(b)$ for each A in $\underline{K\text{-}Aut}_{\mathfrak{C}}^{\cdot}$ satisfying $E^{\cdot}(A) = b$
due to the following lemma:

<u>Lemma</u>: Given a reachable automaton A and an observable
automaton A' in $\underline{K\text{-}Aut}^{\cdot}$ with $E^{\cdot}(A) = E^{\cdot}(A')$ there is a
unique $\underline{K\text{-}Aut}^{\cdot}$-morphism $f:A \to A'$. Moreover f is a reduction iff A' is reachable and $f:S \to S'$ is defined as
being the unique diagonal morphism in:

(9.5.2)

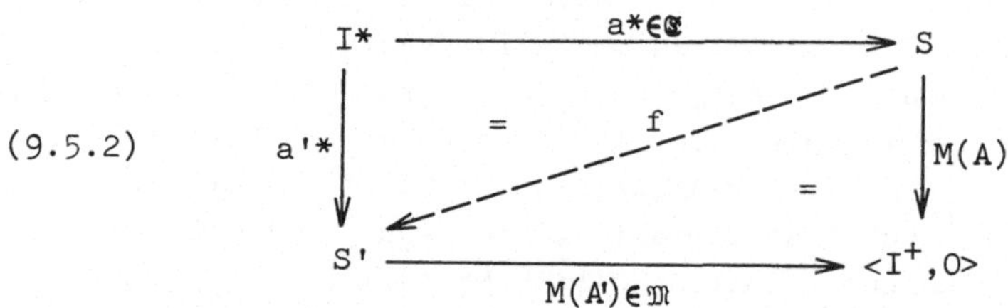

Proof of the lemma: First of all there is a diagonal morphism in (9.5.2) by 4.7,1 because we have $a^* \in \mathfrak{C}$ (A reachable), $M(A') \in \mathfrak{M}$ (A observable) and $M(A) \circ a^* = M(A') \circ a'^*$
which follows from diagram (9.5.3) using the universal prop-
erties of ev_1 (cf. 4.4).

(9.5.3)

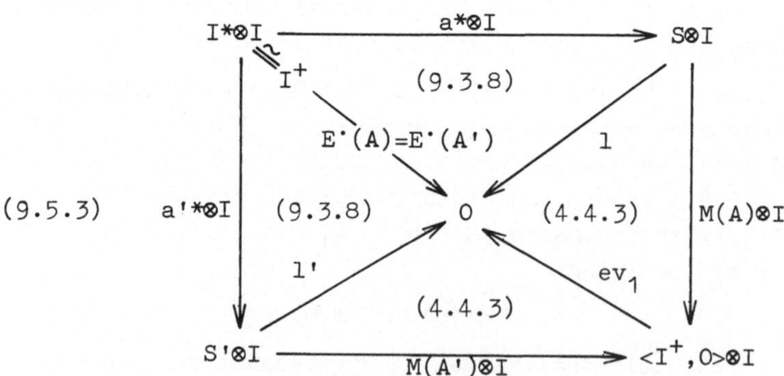

Since M(A) and M(A') are K-Medv-morphisms by 4.4 and
M(A')∘f = M(A) it is easy to show that f:S → S' is also a
K-Medv-morphism using M(A')∈𝔐 . Thus by 4.5 f:V(A) → V(A')
is a K-Aut-morphism but also a K-Aut˙-morphism f:A → A'
because f∘a* = a'* implies f∘a = f∘a*∘u_o = a'*∘u_o = a' by
(9.3.6). Uniqueness of f follows from (9.3.9) and a*∈ℂ .
Finally A' reachable, i.e. a'*∈ℂ , is equivalent to f∈ℂ
by 4.7,3 because we have a'* = f∘a* .   ∎

9.6 Theorem (S t a t e s   o f   t h e   B e h a v i o r
a n d   F i n i t e   R e a l i z a t i o n) : For this
theorem we assume that $(\underline{K},\otimes) = (\underline{K}',\otimes)$ is closed, but the
constructions and results will be carried over to the pseu-
doclosed case in 9.10 .

Given a behavior $b:I^+ \to O$ in B˙ the states S(b) of b
are defined as being the closure of b in $<I^+,O>$ under
left shift L , i.e. S(b) is the image of the unique
K-Medv-morphism $\overline{b}:I* \to <I^+,O>$ in :

(9.6.1)

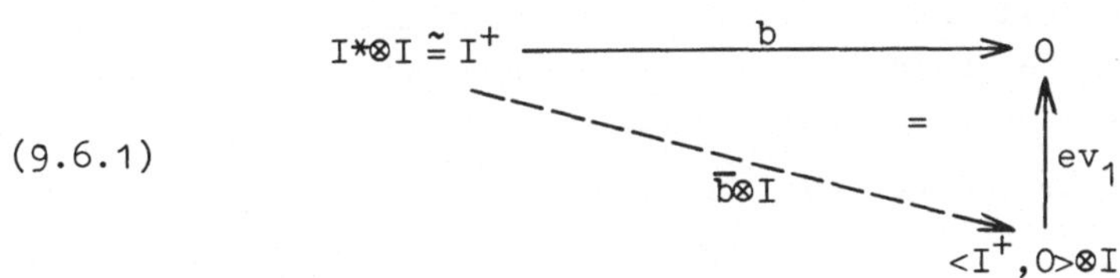

which exists by (4.4.3) with respect to the K-Medv-automata
$I*\otimes I \overset{u}{\to} I*$ (cf. (9.3.2)) and $<I^+,O>\otimes I \overset{L}{\to} <I^+,O>$
(cf. (4.4.1)).
Then we have:

1. S(b) is equal to the (non-initial) behavior EF(b) of
the free realization, i.e. S(b) = EF(b) , and hence to the
state object of the minimal realization M*(b) .

2. Given a class F of "finite" objects in K in the sense

of the remark in 4.8, a behavior b in $\underline{B}^{\cdot}$ is realizable by
a reachable automaton with state object in $\underline{F}$ iff S(b) belongs to $\underline{F}$.

<u>Proof</u>: 1. By the corollary in 4.5 $\bar{b}$ is the machine mor-
phism of the free realization such that the image S(b) of
$\bar{b}$ is equal to the behavior EF(b) of F(b) (cf. (4.6.2))
and hence to the state object of M*(b) by (9.5.1) and theorem 5.5,2.

2. Given A = (S,d,1,a) in $\underline{K}$-$\underline{Aut}^{\cdot}$ with S$\in\underline{F}$ and E$^{\cdot}$(A) = b
we have S(b) = EF(b) by 1. and e(A):S $\rightarrow$ E(A)$\in\mathfrak{C}$ (cf.
(4.6.2)) implies E(A)$\in\underline{F}$ by assumption on $\underline{F}$. Moreover the
reachability morphism a* of A is a $\underline{K}$-$\underline{Aut}^{\cdot}$-morphism
a*:F(b) $\rightarrow$ A by 9.3 and a reduction because A is reachable. Due to 9.4 a* is also a reduction in $\underline{K}$-$\underline{Aut}$ which implies EF(b) = E(A) and hence S(b) = EF(b) = E(A)$\in\underline{F}$. Vice
versa the automaton M*(b) = R$^{\cdot}$F(b) (cf. (9.5.1) has the
state object EF(b) = S(b)$\in\underline{F}$ and realizes b.   ∎

<u>Remark</u>: The reachability assumption in 2 can be avoided
using the next theorem concerning reachability, but we have
to assume that $\underline{F}$ is closed under $\mathfrak{M}$-subobjects in addition,
i.e. f:A $\rightarrow$ B$\in\mathfrak{M}$ and B$\in\underline{F}$ implies A$\in\underline{F}$.

In the following we drop the assumption $(\underline{K},\otimes) = (\underline{K}',\otimes)$ and
consider the more general case of automata in pseudoclosed
categories. However, the reachability construction in 9.7 is
the same in both cases whereas reduction and minimization,
already solved for the closed case in 9.5, is different.

<u>9.7 Theorem</u> (R e a c h a b i l i t y) : For each automa-
ton A there is an equivalent reachable automaton C(A)
and a $\underline{K}$-$\underline{Aut}^{\cdot}$-morphism m:C(A) $\rightarrow$ A with m$\in\mathfrak{M}$ satisfying the
following properties:

1. For each reachable A' and $\underline{K}$-$\underline{Aut}^{\cdot}$-morphism f:A' $\rightarrow$ A
there is a unique f':A' $\rightarrow$ C(A) in $\underline{K}$-$\underline{Aut}^{\cdot}$ satisfying
m$\bullet$f' = f.

2. C(A) is the smallest $\mathfrak{M}$-subautomaton of A, i.e. for each
m':A' → A in K-Aut˙ with m'∈$\mathfrak{M}$ there is a unique
m":C(A) → A' satisfying m'∘m" = m and m"∈$\mathfrak{M}$.

<u>Remark</u>: The couniversal problem in 1 asserts that the in-
clusion functor from the full subcategory of reachable automata has a right adjoint C whereas condition 2 is the corresponding special problem (cf. [23]).
For compatibility of C with reduction and minimization
confer the remarks in 9.8 and 9.9.

<u>Proof</u>: Given A = (S,d,l,a) we construct C(A) = (S̄,d̄,Ī,ā)
in the following way: Let (9.7.1) be a canonical $\mathfrak{E}$-$\mathfrak{M}$-fac-
torization of the reachability morphism a* of A

(9.7.1)

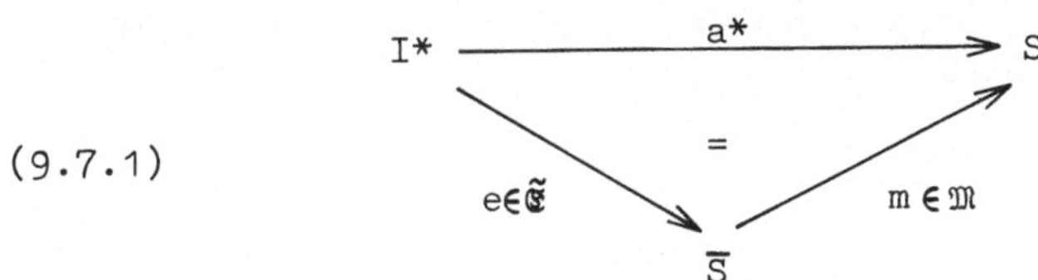

then we define

(9.7.2)
$$\bar{I} := (S̄\otimes I \xrightarrow{m\otimes I} S\otimes I \xrightarrow{1} 0)$$
$$\bar{a} := (U \xrightarrow{u_o} I* \xrightarrow{e} S̄)$$

and d̄:S̄⊗I → S̄ is the unique diagonal morphism in (9.7.3).
Note that (9.7.3) is commutative by (9.7.1) and (9.3.7).

Using (9.7.2) and (9.7.3) and m∘ā = m∘e∘u_o = a*∘u_o = a   it
turns out that m∈$\mathfrak{M}$ is a K-<u>Aut</u>˙-morphism m:C(A) → A and
hence C(A) is equivalent to A .
C(A) is reachable because a* and m are K-<u>Aut</u>˙-morphisms
such that e:I* → S̄ becomes a K-<u>Aut</u>˙-morphism
e:FE˙(A) → C(A) using m∈$\mathfrak{M}$. But this implies that e∈$\tilde{\mathfrak{E}}$ is
the reachability morphism of C(A) by the theorem in 9.3 .
Hence C(A) is reachable.

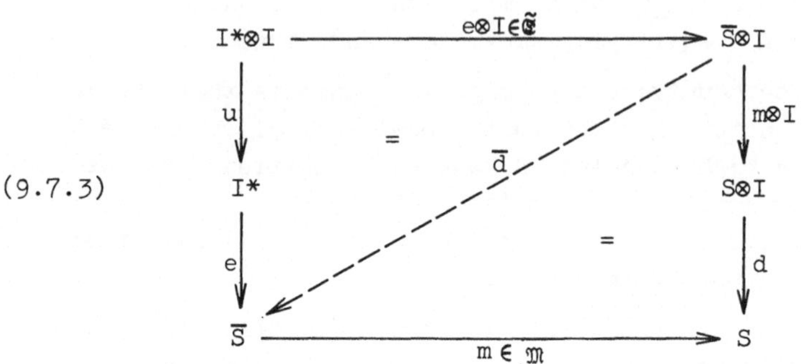

(9.7.3)

In order to show property 1 let $f':S \to \bar{S}$ be the unique diagonal morphism in (9.7.4) which commutes by (9.7.1) and (9.3.9).

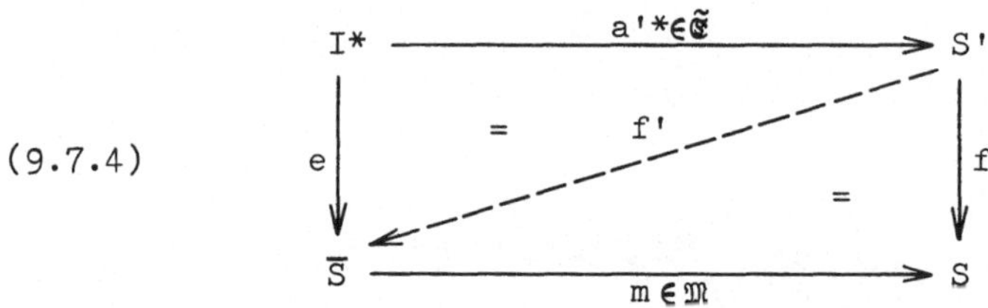

(9.7.4)

Since f and m are <u>K-Aut</u>˙-morphisms and m∘f' = f it is easy to show that f' is a <u>K-Aut</u>˙-morphism and unique using m∈𝔐 .

For the proof of property 2 we take the same diagram (9.7.4) with e∈𝔈̃ and f = m'∈𝔐 such that we get a unique diagonal morphism m":S̄ → S' satisfying m"∘e = a'* and m'∘m" = m .

The same argument as above shows that m"∈𝔐 is a <u>K-Aut</u>˙-morphism.   ∎

<u>Example</u>: Consider the nondeterministic automaton A_2 in (2.7.2). Defining state 5 as initial state in A_2 we obtain an initial automaton $A_2^{\cdot}$ and $C(A_2^{\cdot})$, the reachable subautomaton of $A_2^{\cdot}$, is isomorphic to A_5 in (2.7.5) with initial state b". But if we take 4 as initial state in A_2 we have already a reachable automaton $A_2^{\cdot}{}'$ such that $C(A_2^{\cdot}{}') = A_2^{\cdot}{}'$.

According to the remark in 9.4 we will now show that the construction and the properties of reduced and observable automata in pseudoclosed categories given in 7.3 and 7.4 can be carried over to the initial case.

<u>9.8 Reduction of Initial Automata in Pseudoclosed Categories</u>:
For each $A^{\cdot}$ in $\underline{K}$-$\underline{K}'$-$\underline{Aut}^{\cdot}$ (cf. 9.2) with $V(A^{\cdot}) = A$ the universal reduction $u(A):V(A^{\cdot}) \to R(A)$ given in (7.3.1) can be extended to a reduction $u(A):A^{\cdot} \to R^{\cdot}(A^{\cdot})$ in $\underline{K}$-$\underline{K}'$-$\underline{Aut}^{\cdot}$ by (9.4.2) and (9.4.3) such that $R^{\cdot}(A^{\cdot})$ is reduced and equivalent to $A^{\cdot}$. Moreover given $f:A^{\cdot} \to A^{\cdot}{}'$ in $\underline{K}$-$\underline{K}'$-$\underline{Aut}^{\cdot}$ with reduced $A^{\cdot}{}'$ the unique extension $f':R(A) \to V(A^{\cdot}{}')$ in (7.7.1) is already a $\underline{K}$-$\underline{K}'$-$\underline{Aut}^{\cdot}$-morphism $f':R^{\cdot}(A^{\cdot}) \to A^{\cdot}{}'$ because $\bar{a} = u(A) \circ a$, $a' = f \circ a$ and $f' \circ u(A) = f$ implies $a' = f' \circ \bar{a}$ where $\bar{a}$, a and a' are the initial states of $R^{\cdot}(A^{\cdot})$, $A^{\cdot}$ and $A^{\cdot}{}'$ respectively. Thus we have verified conditions 1 and 2 in 7.7 for $\underline{K}$-$\underline{K}'$-$\underline{Aut}^{\cdot}$ leading to the following extension of theorem 7.7:

<u>Theorem</u>: The subsystematic $\underline{\underline{K}}$-$\underline{\underline{K}}'$-$\underline{\underline{Aut}}^{\cdot}_{red}$ defined by all reduced initial automata is a reduced and realizing subsystematic of $\underline{\underline{K}}$-$\underline{\underline{K}}'$-$\underline{\underline{Aut}}^{\cdot}$ in the sense of 3.3 provided that $\underline{K}'$ has large cointersections and pushouts.

Moreover all the properties given in 7.7,1 to 7.7,6 remain valid in $\underline{K}$-$\underline{K}'$-$\underline{Aut}^{\cdot}$ and reduction preserves reachability using 3.4 and (9.4.4) respectively.

<u>Remark</u>: More precisely reduction is compatible with reachability in the following sense where $R^{\cdot}$ is the reduction and C the reachability functor: For each reduced A in

$\underline{K}$-$\underline{\underline{K}}$'-$\underline{\underline{Aut}}^{\cdot}$ $C(A)$ is reduced and $R^{\cdot}(A')$ is reachable if this is true for A' in $\underline{K}$-$\underline{\underline{K}}$'-$\underline{\underline{Aut}}^{\cdot}$. Furthermore the composite functors $R \circ C$ and $C \circ R$ (restricted to the common domain and codomain) are naturally equivalent (cf. 12.6) .

9.9 Observability Construction for Initial Automata in Pseudoclosed Categories: Similar to 9.8 the weak automata morphism $e(A): A \to A_C$ in 7.8,1 can be extended to a "weak initial automata morphism" $e(A): A^{\cdot} \to A_C^{\cdot}$ for each $A^{\cdot}$ in $\underline{K}$-$\underline{\underline{Aut}}^{\cdot}$ with $V(A^{\cdot}) = A$ by (9.4.2) and (9.4.3) with respect to the following definition:

Definition: Given A and A' in $\underline{K}$-$\underline{\underline{Aut}}^{\cdot}$ a $\underline{\underline{K}}$'-morphism $f: S \to S'$ is called <u>weak initial automata morphism</u> if we have

$$(9.9.1) \qquad\qquad M(A) = M(A') \circ f \qquad \text{and} \qquad f \circ a = a' \, .$$

Similar to 7.6 the corresponding category of initial automata with weak morphisms is denoted by $\underline{K}$-$\underline{\underline{K}}$'-$\underline{\underline{Aut}}^{\cdot W}$ leading to the <u>weak systematic of initial automata</u>
$\underline{\underline{K}}$-$\underline{\underline{\underline{K}}}$'-$\underline{\underline{\underline{Aut}}}^{\cdot W} = (\underline{K}$-$\underline{\underline{K}}$'-$\underline{\underline{Aut}}^{\cdot W}, \underline{\underline{B}}^{\cdot}, E^{\cdot W})$.
$E^{\cdot W}$ is the extension of $E^{\cdot}$ to weak morphisms which is possible because we have $E^{\cdot}(A) = E^{\cdot}(A')$ for each weak morphism $f: A \to A'$ in $\underline{K}$-$\underline{\underline{K}}$'-$\underline{\underline{Aut}}^{\cdot W}$. This is easily seen because (9.9.1) implies (7.6.3) and hence commutativity of (9.2.3).

The same arguments as those given in 9.3 lead to the following weaker version of theorem 7.8 :

Theorem: Provided that each morphism $e \in \mathcal{E}$ has a coretraction c in $\underline{K}$ we have that the subsystematic $\underline{\underline{K}}$-$\underline{\underline{K}}$'-$\underline{\underline{\underline{Aut}}}^{\cdot W}_{obs}$ of the weak systematic $\underline{\underline{K}}$-$\underline{\underline{K}}$'-$\underline{\underline{Aut}}^{\cdot W}$, defined by all observable automata in $(\underline{K}, \otimes)$, is reduced and realizing in the sense of 3.3 but not minimal in general.
Especially we have for each initial automaton $A^{\cdot}$ an equivalent observable automaton $A_C^{\cdot}$ and a weak morphism $e(A^{\cdot}): A^{\cdot} \to A_C^{\cdot}$ which is a reduction in $\underline{K}$-$\underline{\underline{K}}$'-$\underline{\underline{Aut}}^{\cdot}$ iff $A^{\cdot}$

satisfies (7.8.1).

<u>Remark</u>: As before in 7.8 the construction of $A_c^{\cdot}$ depends on the choice of the coretraction c of $e(A)$ but $A_c^{\cdot}$, although observable,is not minimal in $\underline{\underline{K}}$-$\underline{\underline{K}}$'-$\underline{\underline{Aut}}^{\cdot W}$ in general. The corresponding proof in the non-initial case used the fact that $E(A) \subseteq E(A')$ implies a weak morphism $f: A \to A'$ which does not remain true for $E^{\cdot}(A^{\cdot}) = E^{\cdot}(A^{\cdot}{}')$ in the initial case. Hence two equivalent reachable and observable automata do not have isomorphic state objects in general. A counterexample is given below. Finally let us note that the observability of A implies that of $C(A)$, and that for each reachable automaton there is an equivalent reachable and observable automaton in $\underline{\underline{K}}$-$\underline{\underline{K}}$'-$\underline{\underline{Aut}}^{\cdot}$.

<u>Example</u>: The nondeterministic automaton A_5 in (2.7.5) with initial state b'' is reachable and observable but the initial automaton A_8 given in (9.9.2) is an equivalent auto-

(9.9.2)

maton with minimal number of states. Obviously A_8 is also reachable and observable.

<u>9.10 Observable and Finite Realization in the Pseudoclosed Case</u>: Given a behavior $b: I^+ \to O$ in $\underline{\underline{B}}^{\cdot}$ we get a unique $\underline{\underline{K}}'$-morphism $b': I^+ \to PO$ satisfying $v \circ b' = b$ by (6.2.1). Now we construct the minimal realization $M*(b')$ of b' in $\underline{\underline{K}}'$-$\underline{\underline{Aut}}^{\cdot}(I,PO)$ by (9.5.1) which corresponds to an observable automaton $M*(b)$ in $\underline{\underline{K}}$-$\underline{\underline{Aut}}^{\cdot}$ by 8.1. We will show $E^{\cdot}M*(b) = b$ below.

<u>Definition</u>: M*(b) is called <u>observable realization of b</u>.
Similar to 9.6 <u>the states</u> S(b) of b are defined as being
equal to the states S(b') of $b':I^+ \to PO$ in <u>K</u>' as given
in 9.6 for closed categories, i.e. S(b) is the closure of
b' in $<I^+,PO>$ under left shift.

<u>Remark</u>: Another possible construction for M*(b) would be
to take the observable automaton $F(b)_c$ of the free realization F(b) in <u>K-Aut</u>· where c is an arbitrary coretraction of $e(F(b)):I* \to E(F(b))$ (cf. 9.9).

<u>Theorem</u>: 1. M*(b) is an observable realization of b with
"deterministic" state transition, i.e. we have $E^\cdot M*(b) = b$
and the state transition morphism of M*(b) belongs to <u>K</u>'.

2. Given an arbitrary realization A of b , i.e. $E^\cdot(A) = b$,
we have for the (non-initial) behaviors

(9.10.1) $EM*(b) \subseteq E\tilde{P}A$

where $\tilde{P}A$ is the power automaton of A defined in 8.2.

3. S(b) is equal to the state object of the observable realization M*(b) of b.

4. Given a class <u>F</u> of "finite" objects in <u>K</u> satisfying
   (i) K∈<u>F</u> and f:K → L∈ℭ implies L∈<u>F</u> ,
  (ii) L∈<u>F</u> and g:K → L∈𝔐 implies K∈<u>F</u> and
(iii) K∈<u>F</u> implies PK∈<u>F</u>
we have the following characterization of finite behaviors:
A behavior $b:I^+ \to 0$ in <u>B</u>· is realizable by an automaton
with state object in <u>F</u> iff S(b) belongs to <u>F</u>.

<u>Interpretation</u>: 1. M*(b) is the "minimal" realization of b
with respect to automata having a deterministic state transition, but not with respect to arbitrary initial automata
in $(\underline{K},\otimes)$.

2. For nondeterministic and relational automata (9.10.1) implies that for each b'∈EM*(b) there is a set of input-output morphisms in the behavior E(A) of A such that b' is

the union of this set because $E\overset{\circ}{P}A$ is the union closure of
$E(A)$ by 8.3,3. For finite relational automata we get the
following inequality for the cardinality card(S) of states
of an arbitrary realization A:

$$\mathrm{card}(S) \geq \log_2(\mathrm{card}\ S(b))$$

where we have already used part 3 of our theorem. Especially
we have an estimation for the number of states of a "minimal"
realization A of b.

3. We did not define $S(b)$ as being the state object of
$M*(b)$ because we wanted to give a construction of $S(b)$
corresponding to the left shift closure of b, or more precisely of b'.

4. Taking $\underline{F}$ to be finite sets, part 4 of the theorem is
applicable to nondeterministic and relational automata for
example characterizing those behaviors which are realizable
by finite automata. Unfortunately PK is not finite for finite non-empty K in the stochastic case.

Proof of the theorem: 1. By construction of $M*(b)$ it suffices to verify $E^{\cdot}M*(b) = b$. Since we have by 8.1 $1 = v \circ 1'$
for the output morphisms 1 and 1' of $M*(b)$ and $M*(b')$
respectively, we get $E^{\cdot}M*(b) = v \circ E^{\cdot}M*(b')$ using (9.3.8).
Thus we have $E^{\cdot}M*(b) = v \circ E^{\cdot}M*(b') = v \circ b' = b$.

2. By (8.3.3) we have $M(\overset{\circ}{P}F(b)) \circ i_{I*} = M(F(b))$ which implies
$EF(b) \subseteq E\overset{\circ}{P}F(b)$ by (7.6.5). On the other hand $a*:F(b) \to A$
is a $\underline{K}\text{-}\underline{Aut}^{\cdot}$-morphism yielding a $\underline{K}'\text{-}\underline{Aut}(I,PO)$-morphism
$\overset{\circ}{P}(a*):\overset{\circ}{P}F(b) \to \overset{\circ}{P}A$ because $\overset{\circ}{P}$ is a functor by (8.3.1). Hence
we have $E\overset{\circ}{P}F(b) \subseteq E\overset{\circ}{P}A$ and thus $EM*(b) = EF(b) \subseteq E\overset{\circ}{P}F(b) \subseteq E\overset{\circ}{P}A$
using $EM*(b) = EF(b)$ which follows from the construction
of $M*(b)$.

3. We have $S(b) = S(b')$ by definition and $S(b')$ is the state
object of $M*(b')$ by 9.6,1 which is equal to that of $M*(b)$.

4. $S(b) \in \underline{F}$ implies that the state object $S(b)$ of $M*(b)$
belongs to $\underline{F}$. Vice versa, given a realization A of b

with $S \in \underline{F}$ we have by assumption (iii) $PS \in \underline{F}$ and
$e(\tilde{P}A):PS \to E\tilde{P}A \in \mathfrak{C}$ implies $E\tilde{P}A \in \underline{F}$ by (i). Moreover we have
by (9.10.1) and definition of the inclusion $EM*(b) \subseteq E\tilde{P}A$
the existence of a morphism in $\mathfrak{M}$ from $EM*(b)$ to $E\tilde{P}A$
yielding $EM*(b) \in \underline{F}$ and hence $S(b) = EM*(b) \in \underline{F}$ by part 3 of
the theorem. ■

9.11 <u>Initial Power Automata</u>: The construction of power automata given in 8.2 can be extended to the initial case leading to the result that for each initial automaton A in
$(\underline{K}',\otimes)$ the power automaton $\tilde{P} \cdot A$ in $(\underline{K}',\otimes)$ is equivalent to
A. This is a generalization of the well-known equivalence
between deterministic and nondeterministic initial automata.

<u>Theorem</u>: Given an initial automaton $A = (S,d,l,a)$ in the
pseudoclosed category $(\underline{K},\otimes)$ the power automaton
$\tilde{P} \cdot A = (PS,\tilde{d},\tilde{l},\tilde{a})$ of A with $\tilde{d}:PS \otimes I \to PS$ and $\tilde{l}:PS \otimes I \to PO$,
defined by (8.2.1), and $\tilde{a}:U \to PS$, uniquely determined by
(6.2.1) and

(9.11.1) $$v_S \circ \tilde{a} = a \ ,$$

is an initial automaton in the closed category $(\underline{K}',\otimes)$, with
input I and output object PO, which is equivalent to A
in the following sense: $E \cdot (\tilde{P} \cdot A)$ is equal to the $\underline{K}'$-morphism
corresponding to $E \cdot (A)$ by (6.2.1), i.e. the diagram
(9.11.2) commutes:

(9.11.2)

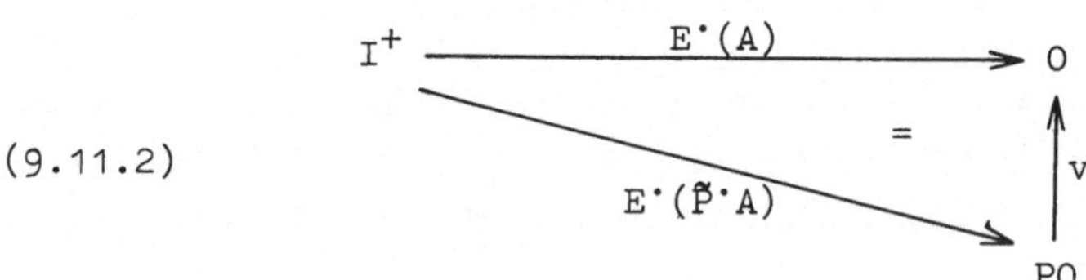

<u>Remark</u>: Similar to 8.3,1 $\tilde{P} \cdot$ becomes a functor
$\tilde{P} \cdot :\underline{K}\text{-}\underline{Aut} \cdot \to \underline{K}'\text{-}\underline{Aut} \cdot (I,PO)$ which is right adjoint to the

inclusion $\mathcal{J}^{\cdot}:\underline{K}'\text{-}\underline{Aut}^{\cdot}(I,PO) \to \underline{K}\text{-}\underline{Aut}^{\cdot}$ where the notation corresponds to that in 8.1 .

<u>Proof</u>: Regarding $\tilde{P}^{\cdot}A$ as an initial automaton $\tilde{P}^{\cdot}A := (PS,\tilde{d},v\circ\tilde{I},\tilde{a})$ in $\underline{K}\text{-}\underline{Aut}^{\cdot}$ v_S becomes a $\underline{K}\text{-}\underline{Aut}^{\cdot}$-morphism $v_S:\tilde{P}^{\cdot}A \to A$ by (8.2.1) and (9.11.1). Thus we have $E^{\cdot}(A) = E^{\cdot}(\tilde{P}^{\cdot}A)$ by (9.2.3), and (9.3.8) yields

$$E^{\cdot}(\tilde{P}^{\cdot}A) = (I^+ \cong I*\otimes I \xrightarrow{\tilde{a}*\otimes I} PS\otimes I \xrightarrow{\tilde{I}} PO \xrightarrow{V} 0) = v\circ E^{\cdot}(\tilde{P}^{\cdot}A)$$

which implies $E^{\cdot}(A) = v\circ E^{\cdot}(\tilde{P}^{\cdot}A)$. The remark is a consequence of the fact that 8.3,1 remains true for initial automata because v_S and f , and hence f' in (8.3.1), are compatible with the initial state morphism. ∎

10. Scoop Minimization

This chapter is a continuation of the last one considering
the minimization problem for initial automata in pseudoclosed categories (cf. 9.2). According to our examples in
9.9 an initial automaton in a pseudoclosed category $(\underline{K},\otimes)$,
which is reachable and observable, is not minimal in general.
Note that this implies minimality in the case of closed categories (cf. 9.5). On the other hand there are no general
constructions for minimizing the number of states for initial stochastic automata for example (cf. [76]), so that we
cannot expect to get such a construction for automata in
pseudoclosed categories. But in most of our examples there
is another construction to decrease the number of states by
replacing a state by an "equivalent subset" of the remaining
states. For nondeterministic automata this means that the
union of the input-output functions of all the states belonging to the subset is equal to the input-output function
of the given state (cf. [75]). In fact, this construction,
called scoop minimization, can be formulated in the framework of automata in pseudoclosed categories and seems to be
a fairly good general approximation for the construction of
initial automata with a minimal number of states.

<u>10.1 General Assumptions</u>: For the first part we make the
same general assumptions as in chapter 9 (cf. 9.1) but in
the second one we need rather strong conditions for our
closed category $(\underline{K}',\otimes)$ which will be introduced in 10.7.

<u>10.2 Example and Motivation</u>: Given the automaton A_9 in
(10.2.1) with initial state 1 we can replace state 3 by the
subset of states $\{1,2\}$ obtaining the automaton A_{10} in
(10.2.2) which is equivalent to A_9.

(10.2.1)

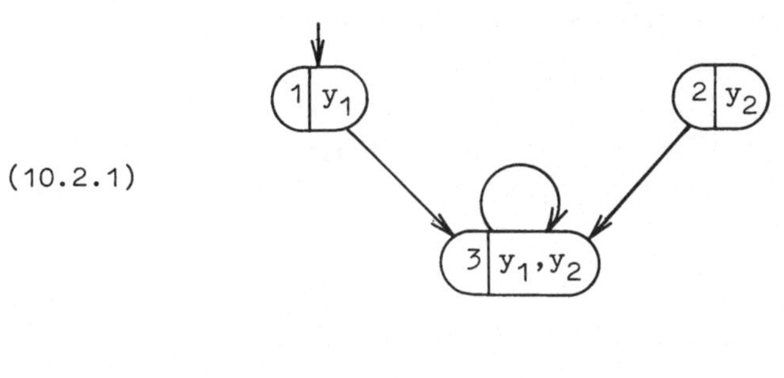

(10.2.2)

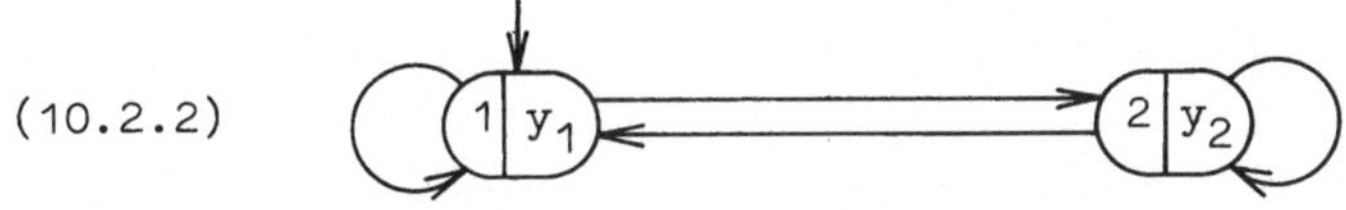

Defining $S' := \{1,2\}$ with inclusion $m:S' \to S$ this replacement is defined by a function $n:S \to PS'$ with $P = P'$
and $n(1) = \{1\}$, $n(2) = \{2\}$ and $n(3) = \{1,2\}$ which satisfies the following equation

$$(10.2.3) \qquad \bigcup_{s' \in n(s)} M(A)(s') = M(A)(s) \qquad \text{for all } s \in S.$$

Using the definition of the union u (cf.6.4) this can be
formulated as a commutative diagram (10.2.4)

$$(10.2.4)$$

$$
\begin{array}{ccc}
S & \xrightarrow{\quad M(A) \quad} & \langle I^+, PO \rangle \\
\downarrow{\scriptstyle n} & = & \uparrow{\scriptstyle u} \\
PS' \xrightarrow[Pm]{} PS & \xrightarrow[PM(A)]{} & P\langle I^+, PO \rangle
\end{array}
$$

where $u \cdot PM(A)$ can be replaced by the machine morphism
$M(\breve{P}A)$ of the power automaton $\breve{P}A$ (cf. (8.3.2)). For the
definition of Pm and PM(A) confer (6.6.2).
Such a pair $(m:S' \to S, n:S \to PS')$ satisfying (10.2.4) will
be called scoop of A because S can be scooped by the

subset S' with respect to the behavior.

<u>10.3 Definition</u> (S c o o p s) : Given an automaton
$A = (S,d,l,a)$ in $(\underline{K},\otimes)$ a pair $(m:S' \to S, n:S \to PS')$ with
$m \in \mathfrak{M}$ and $n \in \underline{K}'$ is called <u>scoop</u> of A if diagram (10.3.1)
is commutative.

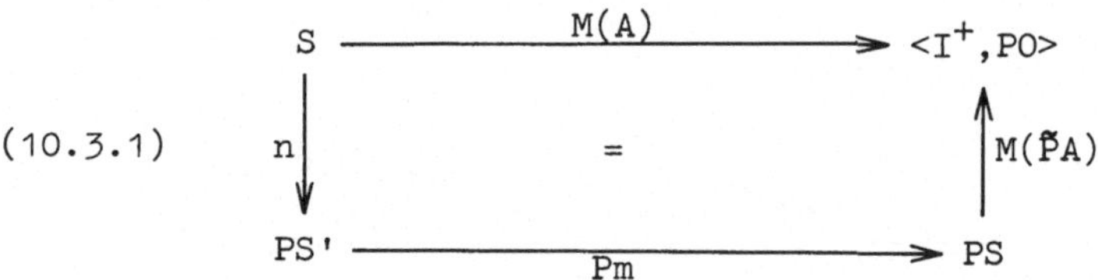

Given such a scoop (m,n) of A the <u>scoop automaton</u>
$A(m,n) = (S',d',l',a')$ of A is defined by the <u>K</u>-morphisms
d', l' and a' in (10.3.2) where $v_{S'}:PS' \to S'$ is the counit of S' (cf. (6.2.1)).

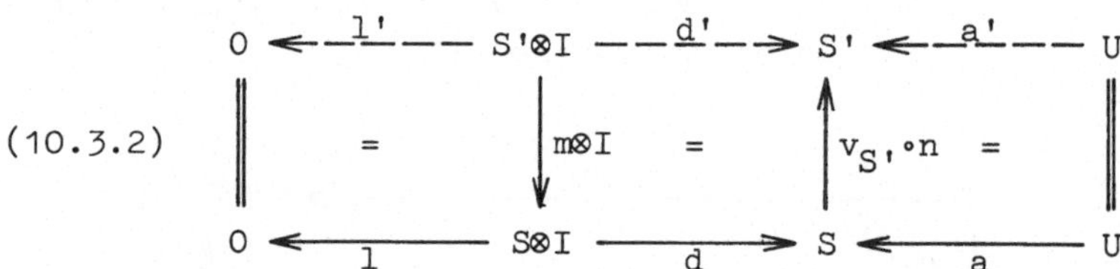

An automaton A is called <u>scoop minimal</u> if for each scoop
(m,n) of A the morphism $m:S' \to S$ is already an isomorphism.

<u>Remark</u>: Although a scoop (m,n) of A does not define an
automata morphism $m:A(m,n) \to A$ in <u>K-Aut</u>· in general, it
follows from diagram (10.3.3) and the following theorem 10.4
that scoops are closed under composition, i.e. given scoops
(m,n) of A and (m',n') of A(m,n) the composition

$(m \circ m', n' * n)$ is a scoop of A satisfying

$$A(m,n)(m',n') = A(m \circ m', n' * n)$$

where $n' * n$ is the composition of n' and n regarded as
$\underline{K}$-morphisms yielding the $\underline{K}'$-morphism $n' * n = P(v_{S''}) \circ P(n') \circ n$.

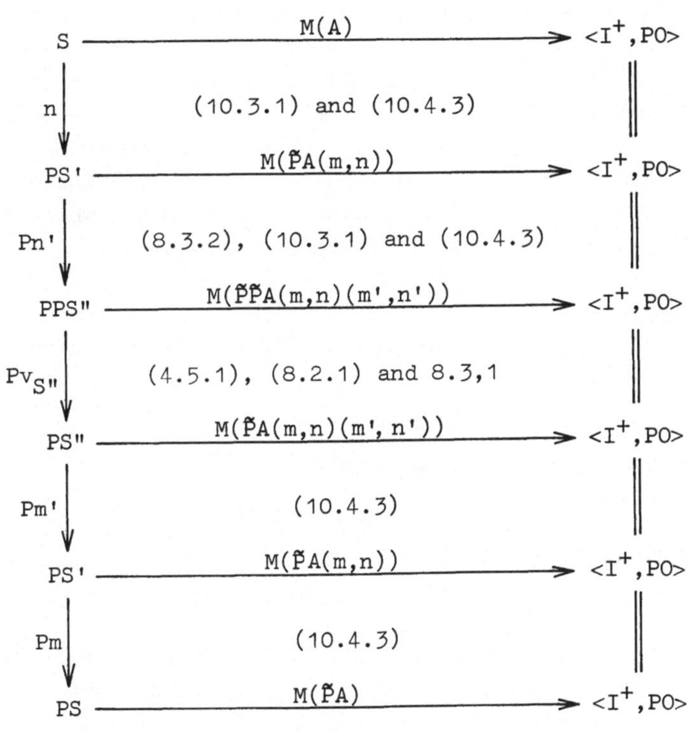

$$(10.3.3)$$

Given an automaton A with "finite" state object S, i.e.
there is no infinite chain of non-trivial subobjects of S,
the composition closure of scoops implies the existence of
one scoop (m,n) of A such that A(m,n) is scoop minimal
and equivalent to A.

In general we will show that for each scoop (m,n) of A the
scoop automaton A(m,n) with state object S' is equivalent to A. Moreover we will give constructions for a scoop
(m,n) and a sufficient condition for A such that A(m,n)
is scoop minimal. But even in this case A(m,n) has not
minimal number of states in general which will be shown in
example 10.6. In 10.2, however, we have given an example for
a scoop (m,n) of the automaton A_9 in (10.2.1) which
leads to a scoop minimal automaton $A_{10} = A_9(m,n)$ which is
also minimal with respect to the number of states.

<u>10.4 Theorem</u> (S c o o p   A u t o m a t a) : Given a scoop
(m,n) of A the scoop automaton A(m,n) is equivalent to
A, i.e.

(10.4.1) $$E^{\cdot}(A(m,n)) = E^{\cdot}(A).$$

Furthermore we have for the machine morphisms

(10.4.2) $$M(A(m,n)) = M(A) \circ m$$

and for the corresponding power automata

(10.4.3) $$M(\tilde{P}A(m,n)) = M(\tilde{P}A) \circ Pm .$$

<u>Proof</u>: In order to prove (10.4.1) we consider the following
diagram (10.4.4) which rows are equal to $E^{\cdot}(A)$ and
$E^{\cdot}(A(m,n))$ respectively. Subdiagram (1) is commutative by
(10.3.2). The commutativity of (2) which is equivalent to
(10.4.2) by (6.3.3) and the following condition (10.4.5)
will be shown below yielding $E^{\cdot}(A) = E^{\cdot}(A(m,n))$ by (10.4.4).

(10.4.5) $$1^{+} \circ (m \otimes I^{+}) \circ ((v_{S'} \circ n) \otimes I^{+}) = 1^{+} .$$

Since (10.4.3) is a direct consequence of (10.4.2) using
(8.3.2) and the functor properties of P it remains to

$$(10.4.4)$$

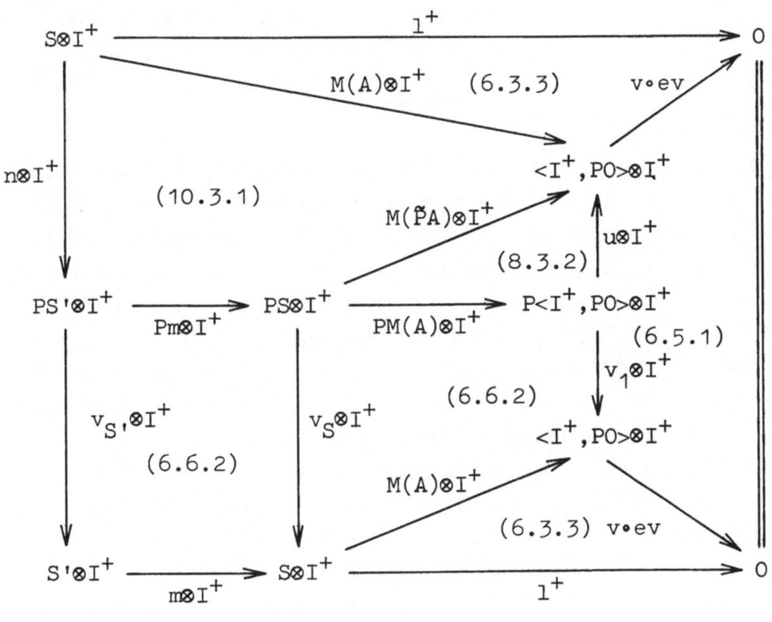

$$(10.4.6)$$

176

verify (2) in (10.4.4) and (10.4.5). (10.4.5) follows from the diagram (10.4.6) above.

Using the coproduct properties of $S \otimes u_k : S \otimes I^k \to S \otimes I^+$ $(k \in \mathbb{N})$ and (6.3.2) the commutativity of (2) in (10.4.4) is equivalent to

$$(10.4.7) \qquad l'_k = l_k \circ (m \otimes I^k) \qquad \text{for all } k \in \mathbb{N}$$

which will be shown by induction. For $k = 1$ this is just the definition of $l' = l'_1$ in (10.3.2) and for $k+1$ we get by (6.3.1), induction hypothesis, (10.3.2), (10.4.5) and (6.3.1) respectively:

$$\begin{aligned}
\underline{l'_{k+1}} &= l'_k \circ (d' \otimes I^k) \\
&= l_k \circ (m \otimes I^k) \circ (d' \otimes I^k) \\
&= l_k \circ (m \otimes I^k) \circ (v_{S'} \circ n) \otimes I^k \circ (d \otimes I^k) \circ (m \otimes I^{k+1}) \\
&= l_k \circ (d \otimes I^k) \circ (m \otimes I^{k+1}) \\
&= \underline{l_{k+1} \circ (m \otimes I^{k+1})}
\end{aligned}$$

$\blacksquare$

<u>10.5 Scoop Construction</u>: Our next step is the construction of a scoop (m,n) for a given automaton A in <u>K-Aut</u>$^{\cdot}$. For this purpose we need in the case of nondeterministic automata the set L of all pairs $(s_i, S_i) \in S \times \wp'S$ such that the state s_i is equivalent to the subset S_i, i.e. $M(A)(s_i) = M(\tilde{\wp}A)(S_i)$. In our general context of automata in pseudoclosed categories the object L together with the projections $p_1 : L \to S$ and $p_2 : L \to PS$ is the "pullback" of the machine morphisms $M(A) : S \to \langle I^+, PO \rangle$ and $M(\tilde{\wp}A) : PS \to \langle I^+, PO \rangle$ which means that we have $M(A) \circ p_1 = M(\tilde{\wp}A) \circ p_2$ in (10.5.1) and for all objects K in $\underline{K}'$ and morphisms $h_1 : K \to S$, $h_2 : K \to PS$ satisfying $M(A) \circ h_1 = M(\tilde{\wp}A) \circ h_2$ there is a unique $h : K \to L$ such that (1) and (2) in (10.5.1) are commutative, i.e. $p_1 \circ h = h_1$ and $p_2 \circ h = h_2$.

For the general definition of pullbacks confer 12.9 .

<u>Assumption</u>: Let $\underline{K}'$ be a category with a pullback for each pair of morphisms $f_1 : A_1 \to B$ and $f_2 : A_2 \to B$.

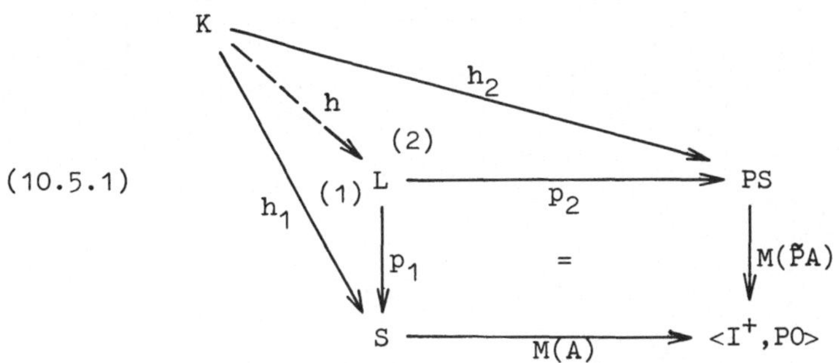

(10.5.1)

Construction: Given an automaton $A = (S,d,l,a)$ in $\underline{K}\text{-}\underline{Aut}^{\bullet}$
we first construct the pullback in (10.5.1) of the machine
morphisms $M(A)$ and $M(\tilde{P}A)$ in $\underline{K}'$. Taking $K = S$, $h_1 = id_S$
and $h_2 = i_S$ in (10.5.1) we have by (8.3.3) $M(\tilde{P}A) \circ i_S = M(A)$
such that there is a unique morphism $h:S \to L$ in (10.5.1)
satisfying $p_2 \bullet h = i_S$ and $p_1 \bullet h = id_S$. Hence p_1 is a retraction in $\underline{K}'$ which has several coretractions $c:S \to L$
in general and h is only one of these. Let c be any coretraction of p_1 , i.e. $p_1 \circ c = id_S$, which is unequal to h
if possible (cf. 10.6). Otherwise our scoop, which depends
on the choice of c , would be trivial. Now we construct an
$\tilde{\mathfrak{E}}\text{-}\mathfrak{m}$-factorization of the $\underline{K}$-morphism $v_S \circ p_2 \bullet c:S \to S$ in
(10.5.2), which exists by assumption (cf. 10.1 and 9.1),
satisfying $m \bullet \tilde{e} = v_S \circ p_2 \circ c$. Moreover, there are unique
$\underline{K}'$-morphisms $n:S \to PS'$ and $Pm:PS' \to PS$ such that (1) and
(2) in (10.5.2) are commutative.
Hence we have $v_S \circ p_2 \circ c = v_S \circ Pm \circ n$ which implies, by uniqueness of f' in (6.2.1),

(10.5.3) $\qquad\qquad\qquad p_2 \circ c = Pm \circ n$.

The pair $(m:S' \to S , n:S \to PS')$ is a scoop of A because
we have by (10.5.3), (10.5.1) and $p_1 \circ c = id_S$

$$M(\tilde{P}A) \circ Pm \circ n = M(\tilde{P}A) \circ p_2 \circ c = M(A) \circ p_1 \circ c = M(A) \quad .$$

178

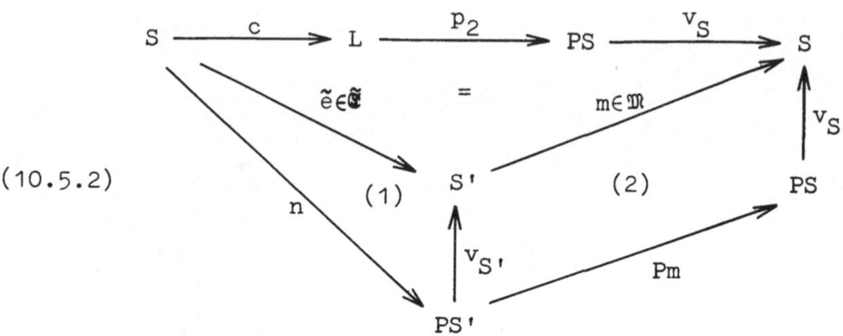

Interpretation: In the case of nondeterministic automata
$p_2 \circ c : S \to P'S$ is a function which assigns to each state s_i
in S an equivalent subset S_i of S , S' is the union of
all these subsets with inclusion $m : S' \to S$, and $n : S \to P'S'$
is the restriction of $p_2 \circ c : S \to P'S$ to $P'S'$. Let us remark
that the choice of the coretraction c of p_1 is important
for the cardinality of S' which is the state set of the
automaton $A(m,n)$. Especially it is useful to assign, as far
as possible, to each state s_i a subset S_i such that s_i
is not an element of S_i . This restriction will be generalized in a special scoop construction in 10.7.

10.6 Example: Given the automaton A_{11} in (10.6.1) with
$I = \{x\}$, $O = \{y_1, \ldots, y_4\}$ and initial state 1 we have the

(10.6.1)

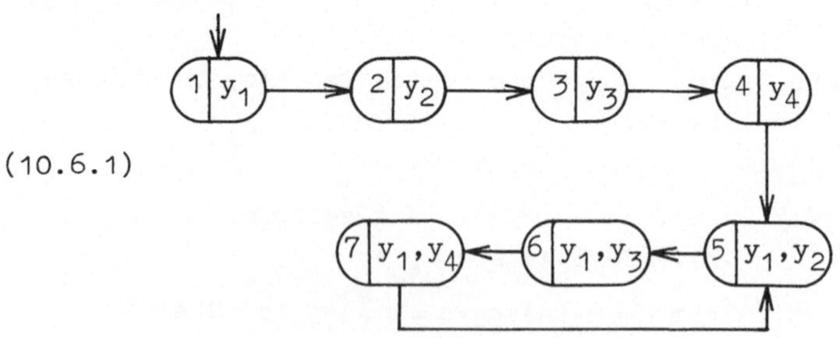

following set L of pairs (s_i, S_i) where $s_i.$ is a state of
A_{11} which is equivalent to the subset S_i of $S = \{1,\ldots,7\}$:

$$L = \{(i,\{i\})/\ i = 1,\ldots,7\} \cup \{(7,\{1,4\})\}\ .$$

The coretraction $h:S \to L$ of $p_1:L \to S$ is given by

$$h(i) = (i,\{i\}) \qquad \text{for}\quad i = 1,\ldots,7$$

but choose the coretraction $c:S \to L$ given by

$$c(i) = h(i) \qquad \text{for}\quad i = 1,\ldots,6 \quad \text{and}\quad c(7) = (7,\{1,4\})$$

such that $S' = \{1,\ldots,6\}$ with inclusion $m:S' \to S$ and
$n:S \to P'S'$ is defined by

$$n(i) = \{i\} \qquad \text{for}\quad i = 1,\ldots,6 \quad \text{and}\quad n(7) = \{1,4\}\ .$$

The scoop automaton $A_{12} = A_{11}(m,n)$ is given in (10.6.2)

(10.6.2)

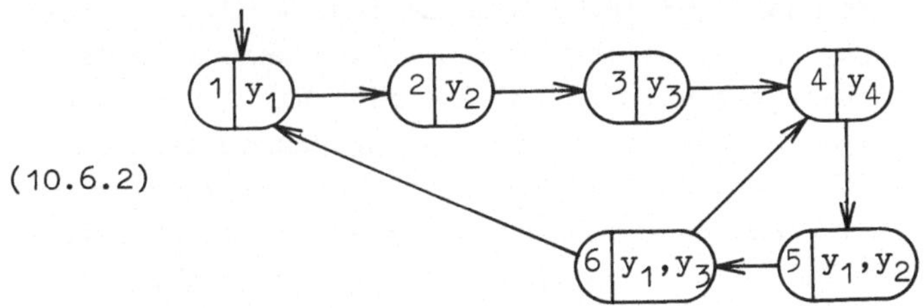

with $d'(6,x) = v_{S'} \circ n \circ d \circ m(6,x) = v_{S'} \circ n \circ d(6,x) = v_{S'} \circ n(7) = \{1,4\}$.

It is easy to check that $A_{11}(m,n)$ is scoop minimal but not
state minimal because A_{13} in (10.6.3) is equivalent to A_{12}
as initial automaton and has only five states.

(10.6.3)

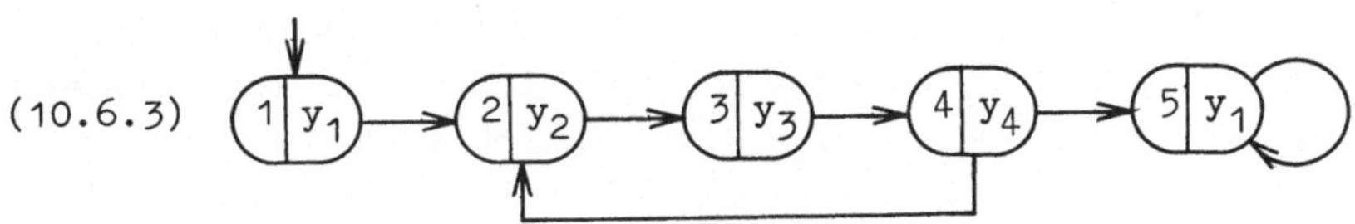

The interesting point in this construction is that $E(\breve{P}A_{12})$ is proper included in $E(\breve{P}A_{13})$ but $E(\breve{P}A_{13})$ can be generated by a smaller number of states. $M(A_{13})(5)$ belongs to $E(\breve{P}A_{13})$ but not to $E(\breve{P}A_{12})$ and we have

$$M(A_{12})(5) = M(A_{13})(2) \cup M(A_{13})(5) = M(\breve{P}A_{13})(\{2,5\})$$
$$M(A_{12})(6) = M(A_{13})(3) \cup M(A_{13})(5) = M(\breve{P}A_{13})(\{3,5\}) \ .$$

<u>Remark</u>: In the case of <u>finite</u> nondeterministic automata a scoop (m,n) of A leading to a scoop minimal automaton $A(m,n)$ can be constructed in the following way: The set $E(\breve{P}A)$ is a semilattice consisting of all the input-output functions $M(A)(s):I^+ \to P'O$ for $s \in S$ and all unions of these functions such that the union of two functions is the union operation in the semilattice. In this finite semilattice $E(\breve{P}A)$ we can determine a minimal subset F of functions generating $E(\breve{P}A)$ with respect to the union: In the first step we take the set F_1 of all minimal elements in $E := E(\breve{P}A)$ and construct the corresponding subsemilattice E_1 of E. In the second step F_2 is the set of minimal elements in $E - E_1$ and E_2 is the subsemilattice of E generated by $F_1 \cup F_2$, etc. We stop the algorithm in the case $E_n = E$ and it terminates because E is finite. In this case $F = F_1 \cup \ldots \cup F_n$ is the minimal generating subset of $E(\breve{P}A)$. Now S' is the set of all states corresponding to the functions in F, $m:S' \to S$ is the inclusion and $n:S \to P'S'$ is given by the semilattice structure. The idea of such a construction is sketched in [6].

Regarding the automaton A_{11} in (10.6.1) $E(\breve{P}A_{11})$ is generated by the functions $f_i = M(A_{11})(i)$ for $i = 1,\ldots,7$ but f_2 is included in f_5 , f_3 in f_6 and the union of f_1 and f_4 is f_7 . Hence the minimal elements in the semilattice $E(\breve{P}A_{11})$ are f_1, f_2, f_3, f_4 such that $F_1 = \{f_1,f_2,f_3,f_4\}$, $f_7 \in E_1$, $F_2 = \{f_5,f_6\}$ is the set of minimal elements in $E - E_1$, and $F = F_1 \cup F_2$ is a minimal subset of functions generating $E(\breve{P}A_{11})$. Since we have $f_1 \cup f_4 = f_7$ and all other functions are generating $n:S \to P'S'$ is de-

fined as above by $n(i) = \{i\}$ for $i = 1,\ldots,6$ and
$n(7) = \{1,4\}$.

The following special scoop construction is more complicated
to formulate and there might be simpler constructions similar to 10.5. But, using this special construction, we can
give a sufficient condition such that the scoop automaton is
(strong) scoop minimal. On the other hand 10.7 and 10.8 may
be omitted without difficulties for further comprehension.

<u>10.7 Special Scoop Construction</u>: As motivated in 10.5 we
give now a more specified construction of a scoop which re-
stricts the choice of the coretraction $c:S \to L$ of $p_1:L \to S$
in the construction 10.5. Moreover we give a sufficient condition for the scoop automaton to be (strong) scoop minimal.
In order to show this theorem we need some additional properties for our closed category $(\underline{K}',\otimes)$, these are satisfied
for the category $(\underline{Set},x)$ but not for $(\underline{Mod}_R,\otimes)$ and $(\underline{Top},\otimes)$
for example. Since $(\underline{ND},x)$, $(\underline{Rel},x)$ and $(\underline{Stoch},x)$ are pseudoclosed relative $(\underline{Set},x)$ and satisfy the additional condition
for P our theorem will at least be applicable to nondeterministic, relational and stochastic automata.

<u>Assumptions</u>: In addition to 10.1 we assume:

A1: $\underline{K}'$ has pullbacks and large intersections (cf. 12.9) and
the composite functor $P \circ J:\underline{K}' \to \underline{K}'$ preserves large intersections.

A2: $\underline{K}'$ has an object G , called <u>generator</u>, which has the
following property: For all pairs $f_1,f_2:A \to B$ of $\underline{K}'$-mor-
phisms we have $f_1 = f_2$ iff $f_1 \circ h = f_2 \circ h$ is satisfied for
all $h:G \to A$ in $\underline{K}'$.

A3: Each morphism $e \in \mathscr{C}$ has a coretraction c in $\underline{K}'$, i.e.
$e \circ c = id$.

A4: Each morphism $m:S' \to S \in \mathfrak{M}$ has a "complement", i.e.
there is an object $(S-S')$ and a morphism $\bar{m}:(S-S') \to S \in \mathfrak{M}$
 such that S is the coproduct of S' and $(S-S')$ with

injections m and $\bar{m}$, and for each $k:G \to S$ in $\underline{K}'$ exactly one of the following two cases is true:

a) there is a $\underline{K}'$-morphism $k_1:G \to S'$ with $m \bullet k_1 = k$

b) there is a $\underline{K}'$-morphism $k_2:G \to (S-S')$ with $\bar{m} \bullet k_2 = k$.

<u>Interpretation for $\underline{K}' = $ Set</u>: 1. The pullback L of two functions $f_i:A_i \to B$ $(i = 1,2)$ is the set of all those pairs $(a_1,a_2) \in A_1 \times A_2$ which satisfy $f_1(a_1) = f_2(a_2)$. A large intersection of injective functions $f_i:A_i \to B$ $(i \in I)$ is the usual intersection of the subobjects $f_i[A_i]$ of B and it is easy to see that $P(\bigcap_{i \in I} A_i) = \bigcap_{i \in I} P(A_i)$ is true for $P(X)$ being the powerset of X with or without empty subset. Moreover we have this condition up to isomorphism if $P(X)$ is the set of all (discrete) probability distributions p on X (cf. 6.2 example 3).

2. Taking $G = \{1\}$ the condition means that two functions $f_1,f_2:A \longrightarrow B$ are equal iff we have $f_1(a) = f_2(a)$ for all $a \in A$ because each $k:G \to A$ is given by the element $k(1) \in A$ and vice versa.

3. For each surjective function $e:A \to B$ the coretraction $c:B \to A$ must satisfy $c(b) \in e^{-1}(b)$ for all $b \in B$ which is possible by the axiom of choice (cf. 7.4).

4. For each subset S' of S there is a complement $S-S'$ and each element $k(1) \in S$ belongs either to S' or to $S-S'$ but not to both of them.

<u>Construction</u>: Given an automaton $A = (S,d,1,a)$ in $\underline{K}\text{-}\underline{Aut}^{\bullet}$ we first construct the pullback in (10.7.1) of $M(A)$ and $M(\tilde{P}A)$ in $\underline{K}'$, as we have done in 10.5, leading to a unique $\underline{K}'$-morphism $h:S \to L$ satisfying $p_2 \circ h = i_S$ and $p_1 \circ h = id_S$ (cf. 10.5). In the example of nondeterministic automata L is the set of all pairs $(s_i,S_i) \in S \times P'S$ such that s_i is equivalent to S_i (cf. 10.5). Clearly s_i can be replaced by S_i in order to decrease the number of states but this only makes sense in the case $S_i \neq \{s_i\}$.

Since $h(s_i) = (s_i, \{s_i\})$ for all $s_i \in S$ we need the complement of all these pairs in L, or more precisely the first
components of the complement which is the set S_1 of all
those elements $s_i \in S$ which are equivalent to at least one
subset S_i unequal to $\{s_i\}$.

(1o.7.1)

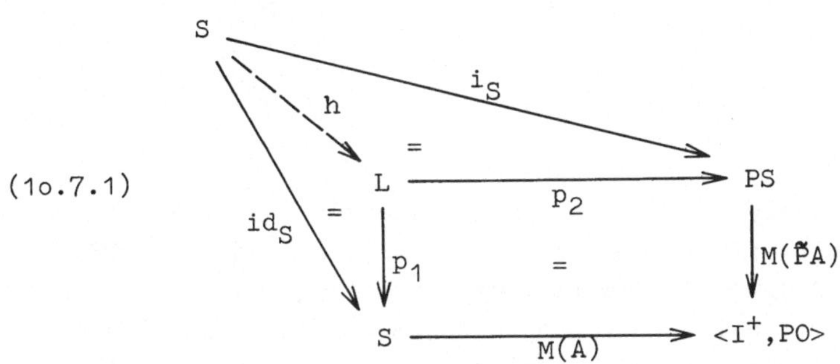

Since h in (10.7.1) is a coretraction and hence in $\mathfrak{M}$ we
have a complement $\bar{h}: (L-S) \to L$ by assumption A4. S_1 will be
defined as the image of the composition $p_1 \cdot \bar{h}$ in (10.7.2).

(10.7.2)

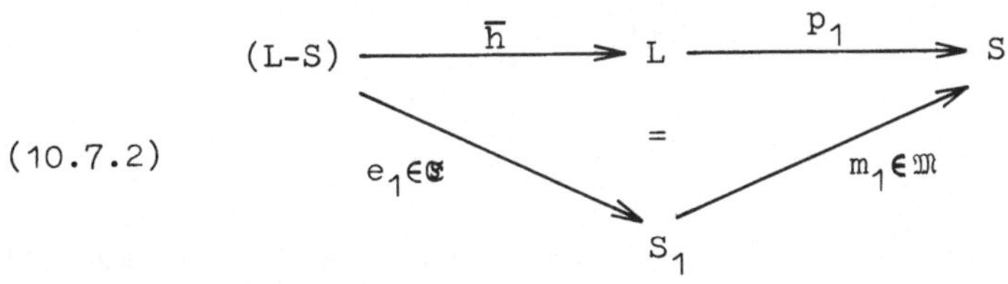

By assumption A1 the $\underline{K}$'-morphism $q := p_2 \cdot \bar{h} \cdot c$ in (10.7.3),
with c being a coretraction of $e_1 \in \mathfrak{E}$, has a factorization
$q = Pm \cdot q'$ where $m: S' \to S$ is the intersection of all "subobjects" $m_i: S_i \to S \in \mathfrak{M}$ admitting such a factorization of q
and containing $(S-S_1)$.

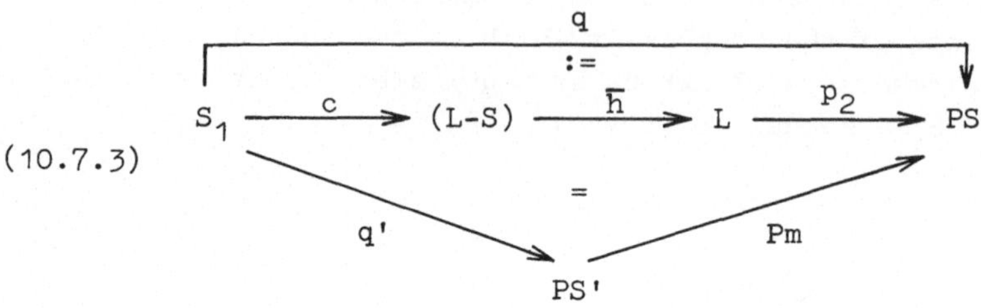

(10.7.3)

More precisely we take the large intersection of all those
morphisms $m_i:S_i \to S \in \mathfrak{M}$ for which there are $q_i^!:S_1 \to PS_i$
and $m_i^!:(S-S_1) \to S_i$ satisfying $Pm_i \circ q_i^! = q$ and $m_i \circ m_i^! = \bar{m}_1$
where $\bar{m}_1:(S-S_1) \to S$ is the complement of $m_1 \in \mathfrak{M}$. Since
$P \circ J$ preserves this intersection we get the factorizations
(10.7.3) and (10.7.4).

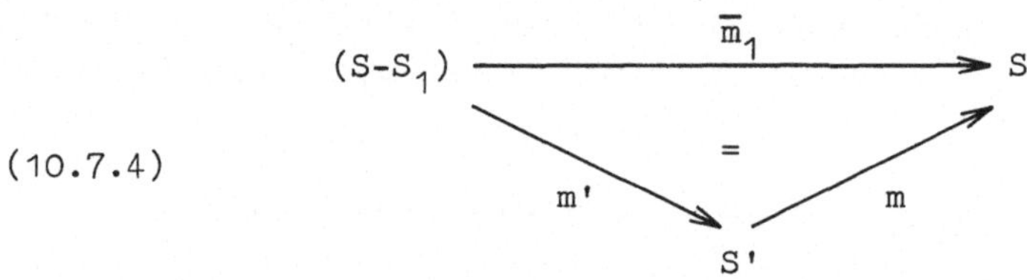

(10.7.4)

In our interpretation q assigns to each $s_1 \in S_1$ a "subset"
$S_1^! \in PS$ which is equivalent to s_1 but unequal to $\{s_1\}$.
S' is the union of $S-S_1$ together with all those subsets
$S_1^! = q(s_1)$ $(s_1 \in S_1)$ and we define $n:S \to PS'$ by $n(s) = S_1^!$ for
$s = s_1 \in S_1$ and $n(s) = \{s\}$ for $s \in (S-S_1)$.
In general $n:S \to PS'$ in (10.7.5) is uniquely defined by
the coproduct properties of m_1 and $\bar{m}_1$ by assumption A4.

The pair (m,n) is a scoop of A which will be shown in
the following theorem.

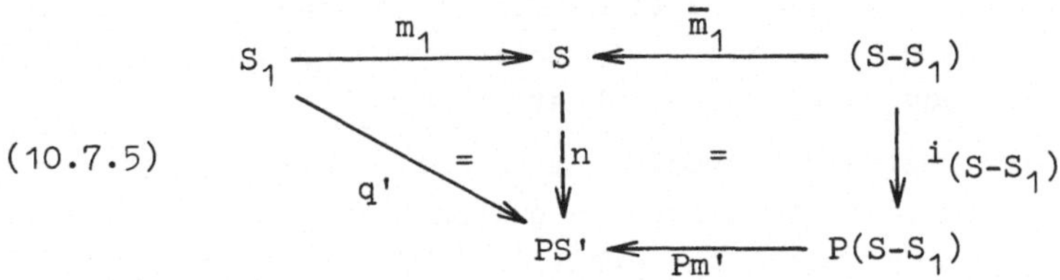

(10.7.5)

10.8 Theorem (S c o o p   M i n i m i z a t i o n) :
Given the assumptions A1 - A4 and the construction in 10.7
the pair $(m:S' \to S, n:S \to PS')$ defined by (10.7.4) and
(10.7.5) is a scoop of A. Moreover the scoop automaton
$A(m,n)$ is scoop minimal if m' in (10.7.4) is an isomorphism, i.e. $(S-S_1) \cong S'$.

Remark: The second part of our theorem is a generalization
of a minimization result in [75] but the concept of scoops
has not been considered in classical automata theory.
In fact we will show in the case $(S-S_1) \cong S'$ that $A(m,n)$
is **strong scoop minimal** in the following sense: For each
pair of morphisms $g:G \to S'$ and $f:G \to PS'$ in (10.8.1)
satisfying $M(A(m,n)) \circ g = M(\tilde{P}A(m,n)) \circ f$ we already have
$i_{S'} \circ g = f$, where G is the generator of assumption A2 .

(10.8.1)

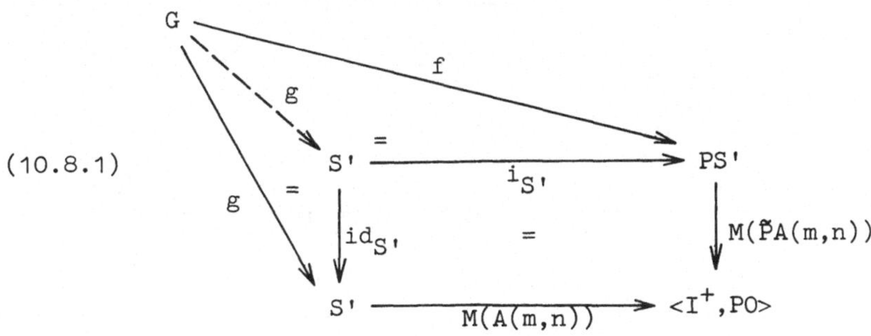

(10.8.1) means for nondeterministic automata that each subset S'_i of S', which is equivalent to a state s_i , is already equal to $\{s_i\}$, i.e. $S'_i = \{s_i\}$, such that no state s_i
in S' is equivalent to a subset $S'_i \neq \{s_i\}$.

If $\underline{K}'$ has arbitrary coproducts it can be shown that the
square in (10.8.1) is already a pullback in $\underline{K}'$, i.e. we have
the universal property for arbitrary objects K in $\underline{K}'$ instead of G.

<u>Proof of the theorem</u>: In order to show that (m,n) is a
scoop of A it suffices to verify

$$(10.8.2) \qquad \begin{aligned} M(A) \circ m_1 &= M(\tilde{P}A) \circ Pm \circ n \circ m_1 \\ M(A) \circ \bar{m}_1 &= M(\tilde{P}A) \circ Pm \circ n \circ \bar{m}_1 \end{aligned}$$

using the universal properties of the coproduct S with injections m_1 and $\bar{m}_1$, and (10.3.1). But (10.8.2) follows
from diagrams (10.8.3) and (10.8.4) below using the fact
that c is coretraction of e_1.

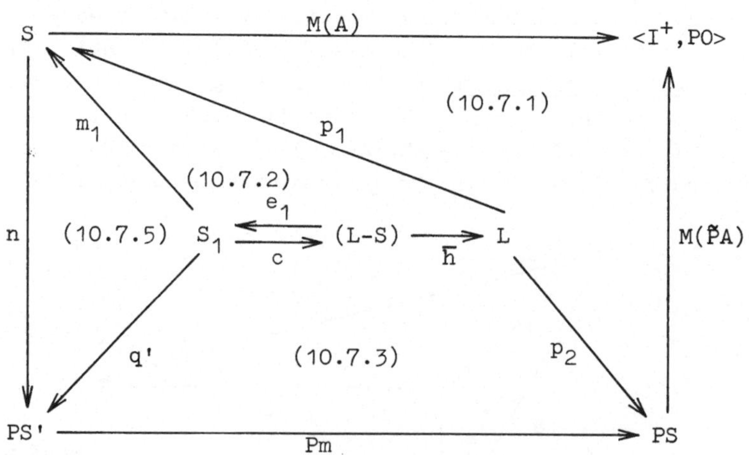

$$(10.8.3)$$

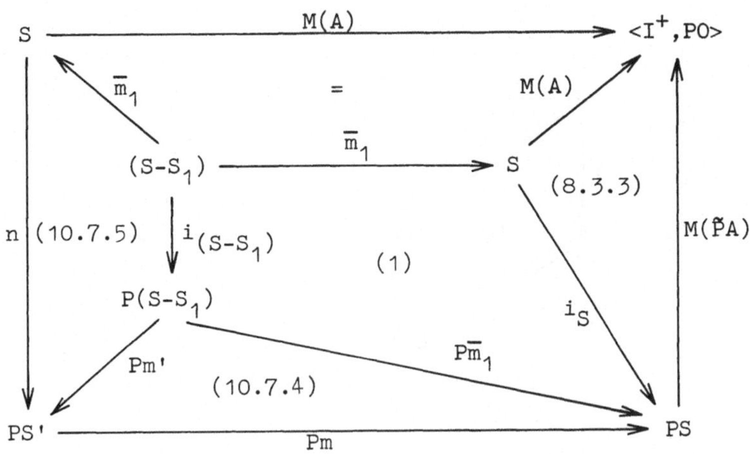

$$(10.8.4)$$

For the commutativity of (1) in (10.8.4) we use the definition of $P\bar{m}_1$ in (6.6.2) and the equations $v_S \circ i_S = id_S$,
$v_{(S-S_1)} \circ i_{(S-S_1)} = id_{(S-S_1)}$ which are the universal definitions of i_S and $i_{(S-S_1)}$ by (6.2.1) respectively
(cf. (8.3.3)).
Now we are going to verify (10.8.1) using assumption A4 in
10.7 and after that it remains to show that (10.8.1) implies
scoop minimality of $A(m,n)$ provided that m' in (10.7.4)
is an isomorphism.

Given $f:G \to PS'$ and $g:G \to S'$ with $M(\tilde{P}A(m,n)) \circ f =$
$= M(A(m,n)) \circ g$ we have to show $i_{S'} \circ g = f$. Using the universal properties of the pullback in (10.8.5) we get a unique
$k:G \to L$ such that diagrams (2) and (3) in (10.8.5) are commutative because we have by (10.4.3) and (10.4.2)

$$M(\tilde{P}A) \circ Pm \circ f = M(\tilde{P}A(m,n)) \circ f = M(A(m,n)) \circ g = M(A) \circ m \circ g \ .$$

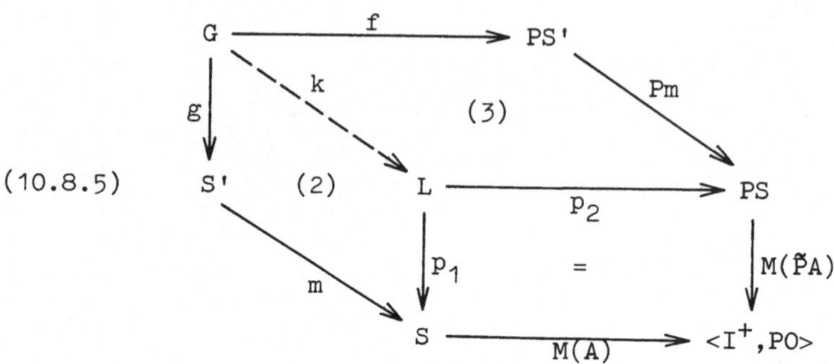

Since Pm is a monomorphism, which can be shown by assumption A1, our equality $i_{S'}\circ g = f$, which has to be shown, is equivalent to $Pm\circ i_{S'}\circ g = Pm\circ f$ and hence to

$$(10.8.6) \qquad i_S\circ p_1\circ k = p_2\circ k$$

using $Pm\circ i_{S'} = i_S\circ m$ and (2), (3) in (10.8.5).

Since $\bar{h}:(L-S) \to L$ is the complement of $h:S \to L$ the morphism $k:G \to L$ can be factored through h or $\bar{h}$ by assumption A4 .

Assuming that there is a $k_1:G \to (L-G)$ with $k = \bar{h}\circ k_1$ we have by (2) in (10.8.5) and (10.7.2)

$$m\circ g = p_1\circ k = p_1\circ \bar{h}\circ k_1 = m_1\circ e_1\circ k_1$$

which contradicts the fact that m_1 is the complement of $\bar{m}_1$ and hence of m because m' in (10.7.4) is an isomorphism by assumption in our theorem. Thus there exists a morphism $k_2:G \to S$ by assumption A4 satisfying

$$(10.8.7) \qquad h\circ k_2 = k$$

which implies (10.8.6) because we have by (10.7.1) :

$$\underline{p_2\circ k} = p_2\circ h\circ k_2 = i_S\circ k_2 = i_S\circ id_S\circ k_2 = i_S\circ p_1\circ h\circ k_2 = \underline{i_S\circ p_1\circ k} \ .$$

But (10.8.6) was the last condition to verify for the strong scoop minimality of $A(m,n)$ in (10.8.1) .

In order to show that A(m,n) is also scoop minimal we
prove the following lemma:

<u>Lemma</u>: Each strong scoop minimal automaton A is also scoop
minimal.

<u>Proof</u>: Let (m,n) be a scoop of A which is strong scoop
minimal and g:G → S be an arbitrary morphism then we have
by (10.3.1)

$$M(\tilde{P}A) \circ Pm \circ n \circ g = M(A) \circ g$$

which implies $i_S \circ g = Pm \circ n \circ g$ because A is strong scoop
minimal. Since g was arbitrary we have by assumption A2 in
10.7 $Pm \circ n = i_S$ and hence by $v_S \circ Pm = m \circ v_{S'}$ and $v_S \circ i_S = id_S$
we obtain $\underline{m \circ v_{S'} \circ n} = v_S \circ Pm \circ n = v_S \circ i_S = \underline{id_S}$ which also implies

$v_{S'} \circ n \circ m = id_{S'}$ because m is a monomorphism. Hence m is
an isomorphism in <u>K</u> showing that A is scoop minimal (cf.
10.3). Note that each isomorphism f in <u>K</u> is also one in
<u>K</u>' because the isomorphisms f and f^{-1} belong to the
class $\mathfrak{M}$ by 4.7,4 , using the $\tilde{\mathfrak{S}}$-$\mathfrak{M}$-factorization in <u>K</u> , and
$\mathfrak{M}$ is a class of morphisms in <u>K</u>' .   ∎

<u>10.9 Remark</u> (M i n i m a l   R e a l i z a t i o n) :
As already mentioned in the introduction to this chapter, a
minimal realization construction similar to that of initial
automata in closed categories (cf. 9.5) we do not hàve in
the pseudoclosed case. A fairly good approximation for a
minimal realization of a behavior $b: I^+ → O \in \underline{B}^{\cdot}$, which can be
constructed in general, seems to be given by the scoop mini-
mization of 10.8 applied to the observable realization M*(b)
in 9.10.
An example for this construction is given in 10.6 where A_{11}
is equal to M*(b) for the function $b: I^+ → P'O$ defined by
$b(x^n) = \{y_n\}$ for n = 1,2,3,4
$b(x^5) = b(x^{5+3k}) = \{y_1, y_2\}$, $b(x^6) = b(x^{6+3k}) = \{y_1, y_3\}$,
$b(x^7) = b(x^{7+3k}) = \{y_1, y_4\}$ for all $k \in \mathbb{N}$.
But a realization of b with minimal number of states is
given by A_{13} in (10.6.3).

11. Structure Theory of Automata

The aim of this chapter is to study the structural properties of automata which are mentioned in 2.9: The construction and characterization of isomorphisms, subautomata,
equalizers, products, coequalizers, coproducts, image-fac-
torizations and free automata respectively. This is already
done in [33] for deterministic automata, and, similar to
that case, it will be shown for automata in monoidal categories $(\underline{K},\otimes)$ that most of the constructions can be "lifted"
from $\underline{K}$, which means that they can be constructed componentwise in the category $\underline{K}$. On the other hand the constructions
of coproducts and free automata in the category of automata
with variable input and output is more difficult and the
proofs are rather long.
We restrict ourselves to Mealy-automata. With slight modifications all results remain true for Moore- and Medvedev-
automata.

11.1 <u>General Assumptions and Remark</u>: In this chapter we only
assume that $(\underline{K},\otimes)$ is a monoidal category.
Let $\underline{K}\text{-}\underline{Aut}$ denote the category of automata in $(\underline{K},\otimes)$ with
fixed I and O (cf. 1.11), and $\underline{K}\text{-}\underline{Aut}_V$ the category of
automata with variable I and O which means that the ob-
jects of $\underline{K}\text{-}\underline{Aut}_V$ are automata (I,O,S,d,l) , and morphisms
between (I,O,S,d,l) and (I',O',S',d',l') are triples
(f_I,f_O,f_S) of $\underline{K}$-morphisms such that (11.1.1) is commutative.

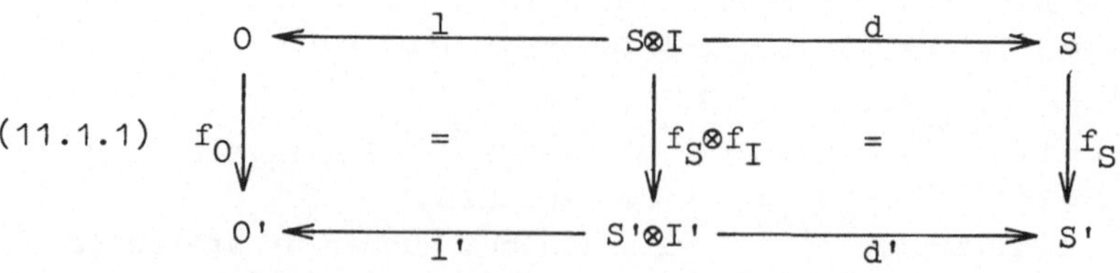

$$(11.1.1)$$

$$191$$

<u>Categorical Remark</u>: Most of our results for the categories
<u>K</u>-<u>Aut</u> and <u>K</u>-<u>Aut</u>$_V$ are corollaries of more general categor-
ical results because, for $(\underline{K},\otimes)$ being closed, <u>K</u>-<u>Aut</u> is
isomorphic to the comma-category $(\underline{K}$-<u>Medv</u> $\downarrow$ $\langle I^+, O\rangle)$ and for
<u>K</u>-<u>Medv</u> and comma-categories (cf. [23]) the constructions
of limits and colimits are well-known.
On the other hand the forgetful functor $V:\underline{K}$-<u>Aut</u>$_V \to \underline{K}^3$,
which assigns to each automaton (I,O,S,d,l) the triplet
(I,O,S) of its objects, is monadic and "algebraic" in the
sense of [65]. Thus it "creates" limits, coequalizers of
V-kernel pairs and, under some restrictions, $\mathfrak{E}$-$\mathfrak{M}$-factoriza-
tions (cf. [71]). Furthermore <u>K</u>-<u>Aut</u>$_V$ has arbitrary colimits
by a theorem of F. E. J. L i n t o n ([58]) provided that
it has coequalizers of reflexive pairs. In 11.5 (ii) we will
construct coproducts without using this restriction.

<u>11.2 Equalizers</u>: Let us consider the category <u>Set</u> :
For given functions $f:M \to M'$, $g:M \to M'$ the set
$\bar{M} = \{a\epsilon M\,/\,f(a) = g(a)\}$ together with the inclusion $j:\bar{M} \to M$
has the properties:

      (i) $f \circ j = g \circ j$ and

      (ii)   for each $h:\tilde{M} \to M$ with $f \circ h = g \circ h$ there is a
            unique $k:\tilde{M} \to \bar{M}$ satisfying $j \circ k = h$.

In categorical language we say that by (i) and (ii) $\bar{M} \overset{j}{\to} M$
is an equalizer of f and g (cf. 12.9) , written also
$Eq \xrightarrow{\;\;eq\;\;} M$ or $Eq(f,g) \xrightarrow{\;\;eq\;\;} M$.

<u>Theorem</u>: Assuming that <u>K</u> has equalizers the categories
<u>K</u>-<u>Aut</u> and <u>K</u>-<u>Aut</u>$_V$ also have equalizers, which can be con-
structed componentwise in <u>K</u> .

<u>Proof</u>: The proof will be given only for <u>K</u>-<u>Aut</u> but it can
be easily carried over to <u>K</u>-<u>Aut</u>$_V$.
Let $A = (S,d,l)$ and $A' = (S',d',l')$ be automata and
$f,g:A \to A'$ a pair of automata morphisms. By assumption
there exists the equalizer $eq:Eq \to S$ of $f:S \to S'$ and
$g:S \to S'$ in <u>K</u> .

192

Now we have

$$f \circ (d \circ (eq \otimes I)) = (f \circ d) \circ (eq \otimes I) = (d' \circ (f \otimes I)) \circ (eq \otimes I) = d' \circ ((f \circ eq) \otimes I)$$
$$= d' \circ ((g \circ eq) \otimes I) = (d' \circ (g \otimes I)) \circ (eq \otimes I) = (g \circ d) \circ (eq \otimes I)$$
$$= g \circ (d \circ (eq \otimes I))$$

and hence by the universal property of the equalizer (Eq, eq)
there exists a unique $\bar{d}: Eq \otimes I \to Eq$ satisfying
$eq \circ \bar{d} = d \circ (eq \otimes I)$. Defining $\bar{I} := 1 \circ (eq \otimes I)$, $\bar{A} = (Eq, \bar{d}, \bar{I})$ becomes an automaton and $eq: Eq \to S$ an automata morphism
from $\bar{A}$ to A . To prove the universal property of $(\bar{A}, eq)$
in $\underline{K}\text{-}\underline{Aut}$ let $h: \tilde{A} = (\tilde{S}, \tilde{d}, \tilde{I}) \to A$ be an automata morphism
with $f \circ h = g \circ h$. By the universal property of (Eq, eq) there
exists a unique $k: \tilde{S} \to Eq$ with $eq \circ k = h$ in (11.2.1) and k
is already an automata morphism from $\tilde{A}$ to $\bar{A}$.

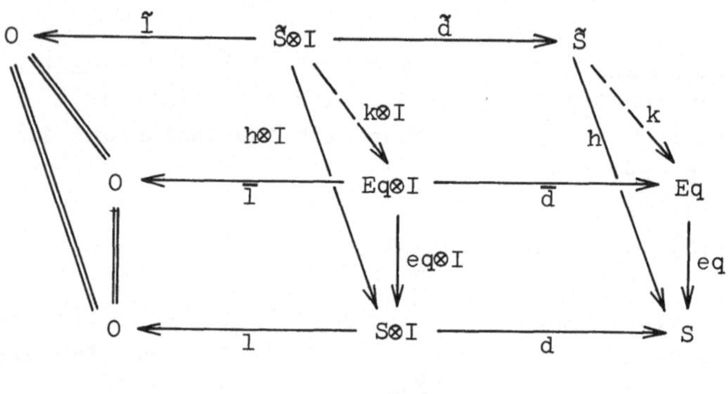

(11.2.1)

In fact we have

$$\bar{I} \circ (k \otimes I) = 1 \circ (eq \otimes I) \circ (k \otimes I) = 1 \circ (h \otimes I) = \tilde{I} \qquad \text{and}$$
$$eq \circ \bar{d} \circ (k \otimes I) = d \circ (eq \otimes I) \circ (k \otimes I) = d \circ (h \otimes I) = eq \circ k \circ \tilde{d}$$

and hence $\bar{d} \circ (k \otimes I) = k \circ \tilde{d}$ by uniqueness of the supplement
morphism into the equalizer Eq .
The uniqueness of k as an automata morphism follows from
the uniqueness as a $\underline{K}$-morphism. ∎

Interpretation: The theorem given above leads to the construction of an equalizer automaton for automata in $(\underline{K},\otimes)$, provided that $\underline{K}$ has equalizers. Hence it is applicable to deterministic, linear, bilinear and topological automata because in $\underline{\text{Mod}}_R$ and $\underline{\text{Top}}$ the equalizers are given in the same way as in $\underline{\text{Set}}$ endowed with submodule structure or subspace topology respectively (cf. [47,48]). Note that it is not applicable to nondeterministic and stochastic automata because $\underline{\text{ND}}$ and $\underline{\text{Stoch}}$ do not have equalizers in general. For partial automata and other types of automata confer 11.6.

11.3 Products: In $\underline{\text{Set}}$ the cartesian product $\underset{i\in J}{X} M_i$ of an arbitrary family $[M_i]_{i\in J}$ has the universal property that each family $[f_i:M \to M_i]_{i\in J}$ of functions induces a unique function $f:M \to \underset{i\in J}{X} M_i$ satisfying $p_j\circ f = f_j$ for all $j\in J$ where p_j denotes the j-th canonical projection.

In an arbitrary category an object together with projections satisfying the universal property given above is called (categorical) product of the family $[M_i]_{i\in J}$ (cf. 12.4), which will be written $\underset{i\in J}{\prod} M_i$.

Now consider a family $[f_i:M_i \to \bar{M}]_{i\in J}$ of functions in $\underline{\text{Set}}$. The set $P:=\{(a_i)_{i\in J} \in \underset{i\in J}{X} M_i \;/\; f_i(a_i)=f_j(a_j) \text{ for all } i,j\in J\}$ together with the projections $[p_j:P \to M_j]_{j\in J}$ has the universal property that

(i) $f_j\circ p_j = f_i\circ p_i$ for all $i,j\in J$

(ii)   for each family $[g_i:\bar{M} \to M_i]_{i\in J}$ of functions with $f_j\circ g_j = f_i\circ g_i$ for all $i,j\in J$ there exists a unique $h:\bar{M} \to P$ satisfying $p_i\circ h = g_i$ for all $i\in J$.

P together with $[p_j]_{j\in J}$ is called fibre product of $[f_i]_{i\in J}$ (cf. 12.9).

A theorem in category theory states that each category has arbitrary fibre products provided that it has arbitrary products and equalizers (cf. 12.9).

<u>Theorem</u> (P r o d u c t s   o f   A u t o m a t a) :

(i) If $\underline{K}$ has products then $\underline{K}\text{-}\underline{Aut}_v$ has products which can be constructed componentwise.

(ii) If $(\underline{K},\otimes)$ is closed and has fibre products then $\underline{K}\text{-}\underline{Aut}$ has products which can be constructed as fibre products in $\underline{K}$.

<u>Proof</u>: (i) Using the same methods as in 11.2 it can be shown that the product automaton of a family $[(I_i,O_i,S_i,d_i,l_i)]_{i \in J}$ can be constructed as $(\underset{i \in J}{\pi} I_i, \underset{i \in J}{\pi} O_i, \underset{i \in J}{\pi} S_i, d, l)$ where d and l in diagram (11.3.1) are induced by the universal properties of the products $\underset{i \in J}{\pi} S_i$ and $\underset{i \in J}{\pi} O_i$ respectively:

$$(11.3.1) \qquad \begin{array}{ccccc} \pi O_i & \xleftarrow{\;\;l\;\;} & (\pi S_i)\otimes(\pi I_i) & \xrightarrow{\;\;d\;\;} & \pi S_i \\ {\scriptstyle p_j^O}\downarrow & = & \downarrow{\scriptstyle p_j^S \otimes p_j^I} & = & \downarrow{\scriptstyle p_j^S} \\ O_j & \xleftarrow{\;\;l_j\;\;} & S_j \otimes I_j & \xrightarrow{\;\;d_j\;\;} & S_j \end{array}$$

(ii) By 4.5 the categories $\underline{K}\text{-}\underline{Aut}$ and $(\underline{K}\text{-}\underline{Medv} \downarrow <I^+,O>)$ are isomorphic. So it suffices to prove that $(\underline{K}\text{-}\underline{Medv} \downarrow <I^+,O>)$ has a product for each family $[m_i : S_i \to <I^+,O>]_{i \in J}$ of $\underline{K}\text{-}\underline{Medv}$-morphisms. Now by assumption $\underline{K}$ has fibre products, and hence $\underline{K}\text{-}\underline{Medv}$ has fibre products which can be proved in an analogous way to the proof in 11.2 : The fibre product in $\underline{K}\text{-}\underline{Medv}$ can be constructed in the S-component.

Obviously the fibre product of the family $[m_i]_{i \in J}$ in $\underline{K}\text{-}\underline{Medv}$ has exactly the same universal property as a categorical product in $(\underline{K}\text{-}\underline{Medv} \downarrow <I^+,O>)$ which completes the proof.

∎

<u>Interpretation</u>: Part (i) of the theorem is applicable to deterministic, linear, bilinear and topological automata because the cartesian product with componentwise defined mod-

ule structure or product topology has the properties of the categorical product in $\underline{Mod}_R$ or $\underline{Top}$ respectively.
For deterministic automata this is exactly the well-known parallel composition of automata (cf. [54]) which means that the state transition and the output function are defined componentwise, e.g. for the state transition function and
$J = \{1,2\}$ $d((s_1,s_2),(x_1,x_2)) = (d_1(s_1,x_1),d_2(s_2,x_2))$.
Note that, replacing "Π" by "X" , the same construction (11.3.1) is possible for nondeterministic, relational and stochastic automata, too, but this construction fails to have the product properties, of course. So in this cases it can be regarded as a tensorproduct in the category of automata. For partial and other automata confer 11.6.

Part (ii) of the theorem is applicable to deterministic, bilinear and topological automata because the corresponding categories $(\underline{Set},x)$, $(\underline{Mod}_R,\otimes)$ and $(\underline{Top},\otimes)$ are closed and have fibre products (cf. above).
The product of a family of observable automata
$[(S_i,d_i,l_i)]_{i \in J}$ has exactly the intersection of the S_i as state object, because observability implies that the machine morphism is a monomorphism.

<u>Corollary</u>: 11.2 and 11.3 together lead (under the given restrictions) to the existence of arbitrary limits in $\underline{K}\text{-}\underline{Aut}_V$ and $\underline{K}\text{-}\underline{Aut}$, especially the existence of pullbacks and hence of intersections of subautomata, invers images of subautomata under automata morphisms and congruence relations generated by automata relations.
For the case $\underline{K} = \underline{Set}$ more details are given in [33].

We will now study the dual concepts of coequalizers and coproducts.

<u>11.4 Coequalizers</u>: In the category of sets the quotient set together with the natural mapping $nat: M \to M_{/\rho}$ induced by an equivalence relation $\rho \subseteq MxM$ has the properties:

(i) $nat \cdot p_1 = nat \cdot p_2$, where p_1 and p_2 are the projections from ρ to M, and

(ii) for each function $g:M \to M'$ with $g \circ p_1 = g \circ p_2$ there is
   a unique function $h:M_{/\rho} \to M'$ satisfying $h \circ nat = g$.
   Define $h([a]) = g(a)$.

These properties lead to the categorical notion of a coequalizer of (p_1,p_2) (cf. 12.9), written
$coeq:M \to Coeq(p_1,p_2)$ or $coeq:M \to Coeq$.

<u>Theorem</u>: (i) If $(\underline{K},\otimes)$ has coequalizers which are preserved
by the functor $-\otimes I$ then <u>K-Aut</u> has coequalizers which can
be constructed in the S-component.

(ii) Assume that $(\underline{K},\otimes)$ has products, large intersections,
coequalizers, a terminal object and kernel pairs and more-
over $-\otimes-$ preserves simultaneously in both variables coequalizers of kernel pairs. Then $\underline{K\text{-}Aut}_v$ has coequalizers
of arbitrary pairs.

<u>Proof</u>: (i) Let $A' = (S',d',1')$ and $A = (S,d,1)$ be two au-
tomata and $f,g:S' \to S$ two automata morphisms in <u>K-Aut</u> from
A' to A. By assumption there exists $coeq:S \to Coeq(f,g)$ in
$\underline{K}$ and $coeq \otimes I:S \otimes I \to Coeq(f,g) \otimes I$ is the coequalizer of $f \otimes I$
and $g \otimes I$. Now we have

$$(coeq \circ d) \circ (f \otimes I) = coeq \circ (d \circ (f \otimes I)) = coeq \circ (f \circ d') = coeq \circ (g \circ d')$$
$$= coeq \circ (d \circ (g \otimes I)) = (coeq \circ d) \bullet (g \otimes I)$$

and hence by the universal property of the coequalizer there
exists a unique $\bar{d}:Coeq \otimes I \to Coeq$ and similarly a unique
$\bar{1}:Coeq \otimes I \to 0$ such that (11.4.1) is commutative

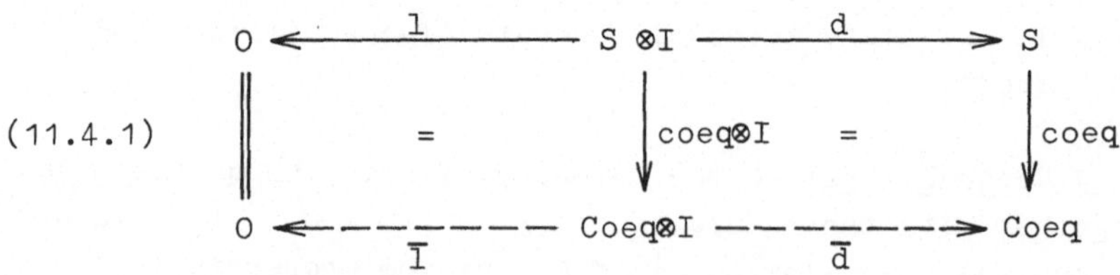

(11.4.1)

which means that $(\text{Coeq},\bar{d},\bar{I})$ is an automaton and coeq an automata morphism.

In an analogous way to the proof of 11.2 it can be verified that $(\text{Coeq},\bar{d},\bar{I})$ has the universal properties of a coequalizer in $\underline{\text{K-Aut}}$.

(ii) In analogy to (i) it can be shown that the coequalizer of a given kernel pair $(f:A' \to A, g:A' \to A) =$
$= ((f_I,g_I),(f_O,g_O),(f_S,g_S))$ in $\underline{\text{K-Aut}}_v$ can be constructed by $(\text{Coeq}^I,\text{Coeq}^O,\text{Coeq}^S,\bar{d},\bar{I})$ where $\text{Coeq}^I := \text{Coeq}(f_I,g_I)$ etc. and $\bar{d}$ and $\bar{I}$ are induced by the universal property of the coequalizer $\text{Coeq}^S \otimes \text{Coeq}^I$ in (11.4.2).

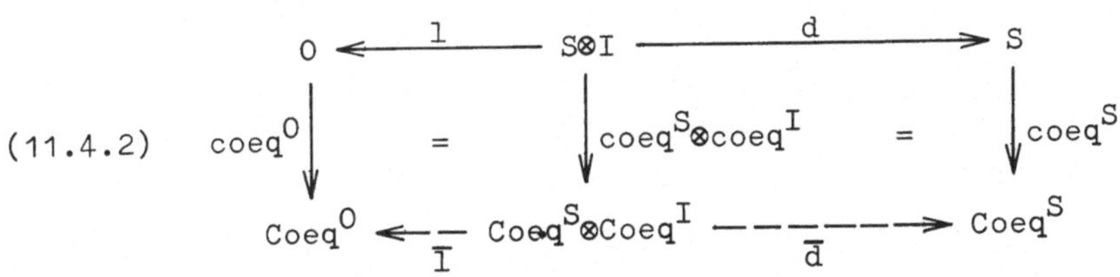

$$(11.4.2)$$

Hence by a general theorem in category theory (cf. [33], 3.53) the given assumptions lead to the existence of coequalizers of arbitrary pairs which can be constructed as the coequalizer of the generated kernel pair. For the construction of this kernel pair we only need large intersections (cf. 12.9) and not necessarily arbitrary fibre products in a well-powered category which is used in [33]. ∎

<u>Interpretation</u>: Part (i) of the theorem is applicable to deterministic, linear, bilinear and topological automata. The coequalizer of an arbitrary pair of functions $f,g:M' \to M$ in $\underline{\text{Set}}$ can be constructed by $\text{nat}:M \to M_{/\bar{\rho}}$ where $\bar{\rho}$ is the minimal equivalence relation on M containing $\rho = \{(f(a),g(a))/\ a \in M'\} \subseteq M \times M$. The same construction gives the coequalizer in $\underline{\text{Mod}}_R$ and $\underline{\text{Top}}$, endowed with quotient module structure or quotient topology respectively (cf.[47,48]).

Note that in the deterministic, bilinear and topological case the functor $-\otimes I$ preserves coequalizers because these categories are closed and hence $-\otimes I$ is left adjoint (cf. 12.9 remark 3). The same is true for linear automata because $-\times I$ preserves quotients in $\underline{Mod}_R$. For nondeterministic and stochastic automata we have the same difficulty as in 11.2 for equalizers, $\underline{ND}$ and $\underline{Stoch}$ do not have coequalizers in general.

For partial and other automata confer 11.6.

Part (ii) of the theorem is applicable e.g. to automata in categories which are "cartesian closed", i.e. "closed" with respect to the categorical product as a tensorproduct. Moreover it is applicable to partial automata.

<u>11.5 Coproducts</u>: As already mentioned in 4.2 the disjoint union of sets leads to the categorical notion of a coproduct (cf. 12.4).

<u>Theorem</u> (C o p r o d u c t s   o f   A u t o m a t a) : We assume that $(\underline{K},\otimes)$ has coproducts which are preserved by $-\otimes I$ and $I\otimes-$ for all $I\in\underline{K}$.

(i) Then $\underline{K}$-<u>Aut</u> has coproducts which can be constructed in the S-component.

(ii) Then $\underline{K}$-<u>Aut</u>$_V$ has coproducts. Here the I-component of the coproduct is the coproduct of the I-components and the constructions of the S- and O-components, which are more difficult, are given in the proof.

<u>Proof of (i)</u>: Let $[(S_i,d_i,1_i)]_{i\in J}$ be a given family of automata. There exists the coproduct $\coprod\limits_{i\in J} S_i$ in $\underline{K}$ together with the family of injections $j_k\colon S_k \to \coprod\limits_{i\in J} S_i$ and by assumption $(\coprod\limits_{i\in J} S_i)\otimes I$ is a coproduct of $[S_i\otimes I]_{i\in J}$ with injections $j_k\otimes I\colon S_k\otimes I \to (\coprod\limits_{i\in J} S_i)\otimes I$.

Using the universal property of the coproduct $(\coprod_{i \in J} S_i) \otimes I$
there exist unique d and l such that (11.5.1) is commutative for all $k \in J$.

$$(11.5.1) \qquad \begin{array}{ccccc} 0 & \xleftarrow{\quad l \quad} & (\coprod S_i) \otimes I & \xrightarrow{\quad d \quad} & \coprod S_i \\[2mm] \| & {\scriptstyle =} & \uparrow{\scriptstyle j_k \otimes I} & {\scriptstyle =} & \uparrow{\scriptstyle j_k} \\[2mm] 0 & \xleftarrow[\quad l_k \quad] & S_k \otimes I & \xrightarrow[\quad d_k \quad] & S_k \end{array}$$

Analogous to the proof in 11.2 it can be verified that
$(\coprod_{i \in J} S_i, d, l)$ has the universal property of a coproduct in
$\underline{K}\text{-}\underline{Aut}$. $\qquad\qquad\qquad\qquad\qquad\qquad\qquad\blacksquare$

<u>Proof of (ii)</u>: The construction of a coproduct in $\underline{K}\text{-}\underline{Aut}_v$ is
much more complicated and it seems to be rather "uncanonical".
We will show that the coproduct of a given family
$[A_i = (I_i, O_i, S_i, d_i, l_i)]_{i \in J}$ can be constructed by
$A = (I, O, S, d, l)$ with

$$I := \coprod_{i \in J} I_i \quad , \quad S := (\coprod_{i \in J} S_i) \coprod D \quad , \quad O := (\coprod_{i \in J} O_i) \coprod D \quad \text{and}$$

$$D := \coprod_{\substack{(i,j) \in J^2 \\ i \neq j}} (S_i \otimes I_j \otimes I^*).$$

To get d and l we first construct an isomorphism
$q : S \otimes I \xrightarrow{\sim} \coprod_{i \in J} (S_i \otimes I_i) \coprod D$ defined by the diagram (11.5.2),
where I^* denotes the free monoid over I (cf. (9.3.1)).
In <u>Set</u> for example q is elementwise the identity, i.e.
$q(x) = x$ for all x where we have only changed the coproduct component.
By construction q is compatible with all coproduct injections.

$$S \otimes I = [(\coprod_i S_i) \amalg (\coprod_{i \neq j} S_i \otimes I_j \otimes (\coprod_k I_k)^*)] \otimes [\coprod_k I_k]$$

$$\downarrow \simeq$$

$$[\coprod_i S_i \otimes \coprod_k I_k] \amalg [\coprod_{i \neq j} S_i \otimes I_j \otimes (\coprod_k I_k)^* \otimes \coprod_k I_k]$$

$$q \downarrow \qquad := \qquad \downarrow \simeq$$

$$[\coprod_i S_i \otimes I_i] \amalg [\coprod_{i \neq j} S_i \otimes I_j] \amalg [\coprod_{i \neq j} S_i \otimes I_j \otimes (\coprod_k I_k)^+]$$

$$\downarrow \simeq$$

$$(\coprod_i S_i \otimes I_i) \amalg D = [\coprod_i S_i \otimes I_i] \amalg [\coprod_{i \neq j} S_i \otimes I_j \otimes (\coprod_k I_k)^*]$$

$$(11.5.2)$$

d and l are now defined by diagram (11.5.3)

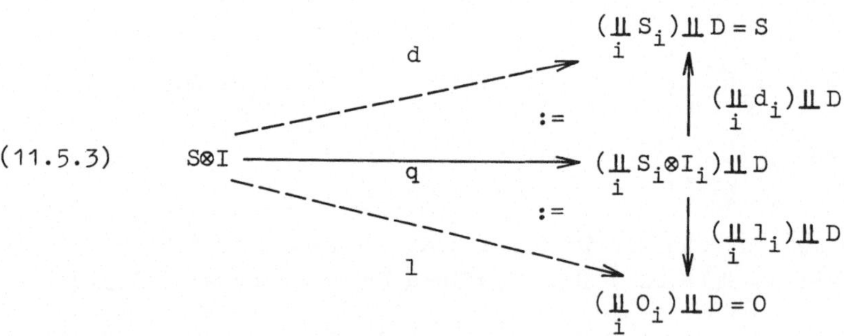

For the definition of $\coprod_i d_i$ for example, confer subdiagram (1) in (11.5.5) where $\tilde{\jmath}_k : S_k \otimes I_k \to \coprod_i (S_i \otimes I_i)$ are the coproduct injections.

The injections $\mathcal{J}_k = (\mathcal{J}_k^I, \mathcal{J}_k^O, \mathcal{J}_k^S): A_k \to A$ are defined by the equations (11.5.4) where all morphisms are coproduct injections:

$$\mathcal{J}_k^I := (I_k \xrightarrow{\ j_k^I\ } \underset{i}{\amalg} I_i = I)$$

$$(11.5.4) \quad \mathcal{J}_k^O := (O_k \xrightarrow{\ j_k^O\ } \underset{i}{\amalg} O_i \xrightarrow{\ j_1^O\ } (\underset{i}{\amalg} O_i) \amalg D = O)$$

$$\mathcal{J}_k^S := (S_k \xrightarrow{\ j_k^S\ } \underset{i}{\amalg} S_i \xrightarrow{\ j_1^S\ } (\underset{i}{\amalg} S_i) \amalg D = S) \quad ,$$

and together with a similar diagram for the output functions (11.5.5) shows that $\mathcal{J}_k: A_k \to A$ is an automata morphism for each $k \in J$:

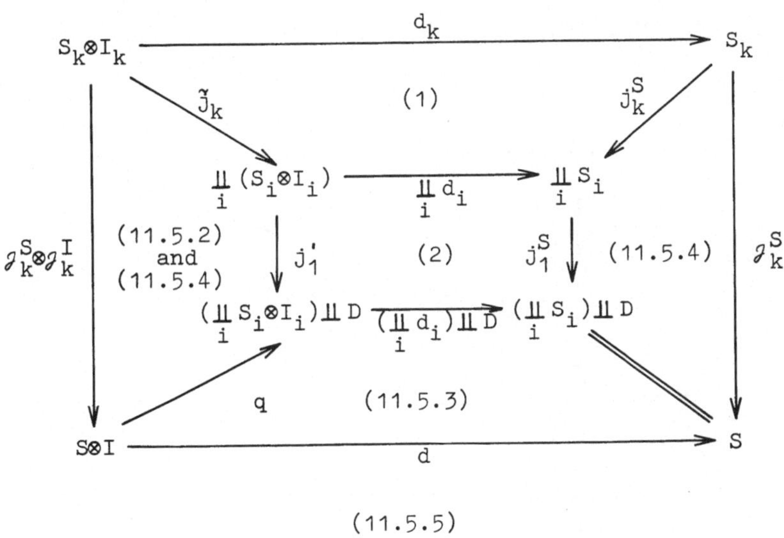

$$(11.5.5)$$

The subdiagrams (1) and (2) are commutative by definition of $\underset{i}{\amalg} d_i$ and $(\underset{i}{\amalg} d_i) \amalg D$ respectively.

So it remains to prove that $A = (I, O, S, d, l)$ has the universal property of a coproduct in $\underline{K\text{-}Aut}_V$.

Let us assume that $[f_i = (f_i^I, f_i^O, f_i^S)]_{i \in J}$ is a family of automata morphisms, $f_i : A_i \to A' = (I', O', S', d'l')$, and that there exists an automata morphism $f = (f^I, f^O, f^S) : A \to A'$ satisfying $f \circ \mathcal{J}_k = f_k$, i.e.

$$f^I \circ \mathcal{J}_k^I = f_k^I$$

(11.5.6) $\quad f^O \circ \mathcal{J}_k^O = f_k^O \qquad$ for all $\quad k \in J$

$$f^S \circ \mathcal{J}_k^S = f_k^S$$

We will show that there exists at most one f with (11.5.6). The existence of such an f will be shown later on.

From (11.5.6) there follow directly the equations (11.5.7) – (11.5.9):

(11.5.7) $\quad f^I = \langle f_i^I \rangle_{i \in J} : \coprod_i I_i \to I'$, i.e. f^I is uniquely determined by the f_k^I .

(11.5.8)

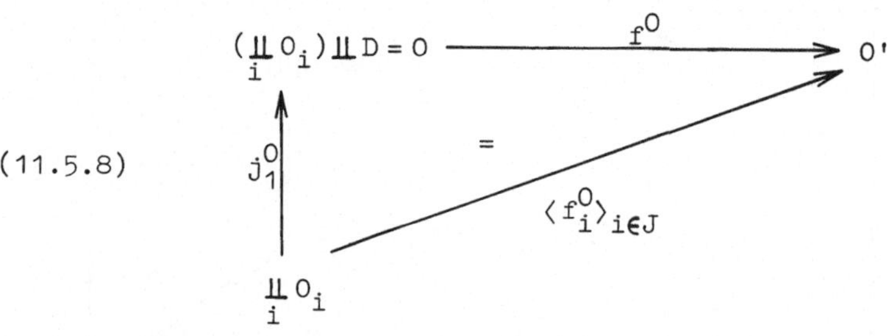

(11.5.9)

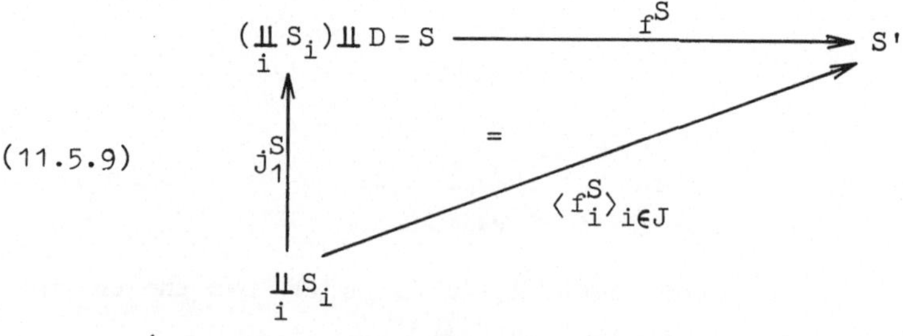

So it remains to consider

$$D \xrightarrow{\;j_2^O\;} (\coprod_i O_i) \amalg D = O \xrightarrow{\;f^O\;} O' \qquad \text{and}$$

$$D \xrightarrow{\;j_2^S\;} (\coprod_i S_i) \amalg D = S \xrightarrow{\;f^S\;} S'$$

which will be characterized in (11.5.13) and (11.5.12) respectively.

Using the definition of $I*$ in (9.3.1) the isomorphism

$$D = \coprod_{i \neq k} S_i \otimes I_k \otimes (\coprod_j I_j)* \;\cong\; \coprod_{n \geq 0} \coprod_{i \neq k} \coprod_{(k_1,\ldots,k_n) \in J^n} S_i \otimes I_k \otimes I_{k_1} \otimes \ldots \otimes I_{k_n}$$

yields a family

$$[\, \jmath_{(i,k,k_1,\ldots,k_n)} : S_i \otimes I_k \otimes I_{k_1} \otimes \ldots \otimes I_{k_n} \to D\,]_{\substack{n \geq 0 \\ (i,k,k_1,\ldots,k_n) \in J^{n+2} \\ i \neq k}}$$

of coproduct injections. Below we will prove that the following diagram (11.5.10) commutes for all $(n+2)$-tuples $(i,k,k_1,\ldots,k_n)$ with $i \neq k$

$$
\begin{array}{ccccc}
S_i \otimes I_k \otimes I_{k_1} \otimes \ldots \otimes I_{k_n} & \xrightarrow{\;\jmath_{(i,k,k_1,\ldots,k_n)}\;} & D & \xrightarrow{\;j_2^S\;} & (\coprod_i S_i) \amalg D \\[4pt]
\Big\downarrow{\scriptstyle f_i^S \otimes f_k^I \otimes f_{k_1}^I \otimes \ldots \otimes f_{k_n}^I} & = & & & \Big\| \; S \\[4pt]
& & & & \Big\downarrow{\scriptstyle f^S} \\[4pt]
S' \otimes I'^{\otimes I'^n} & \xrightarrow{\hspace{4cm} d'_{n+1} \hspace{4cm}} & & & S'
\end{array}
$$

$$(11.5.10)$$

where the family $[\,d'_{n+1} : S' \otimes I'^{n+1} \to S'\,]_{n \geq 0}$ is recursively defined by (11.5.11) and the following equation:

$$d'_1 = d : S' \otimes I' \to S'$$

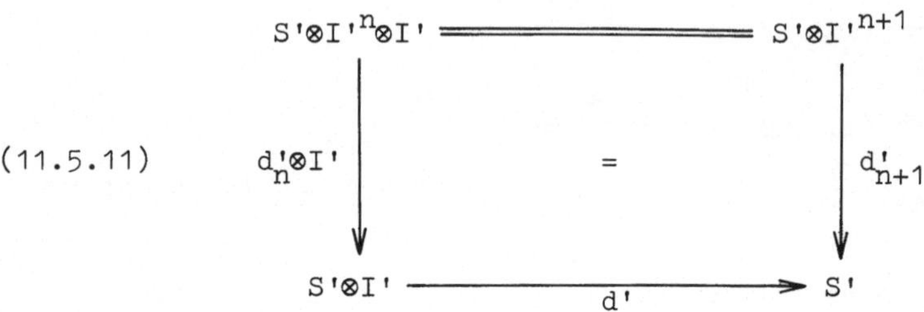

(11.5.11)

Using (11.5.10) we have the commutativity of (11.5.12).

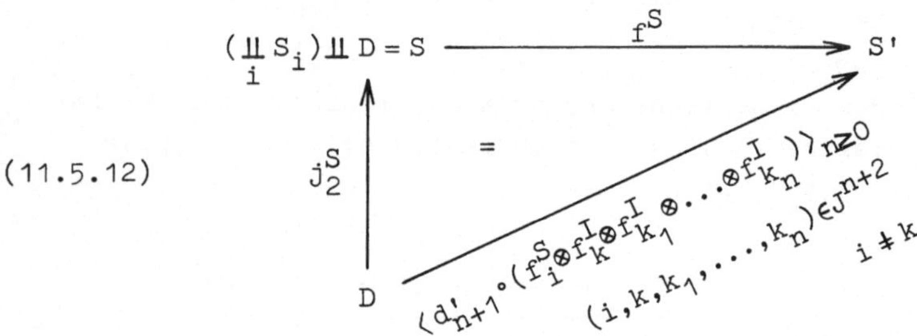

(11.5.12)

By similar methods it can be shown that (11.5.13) is commutative

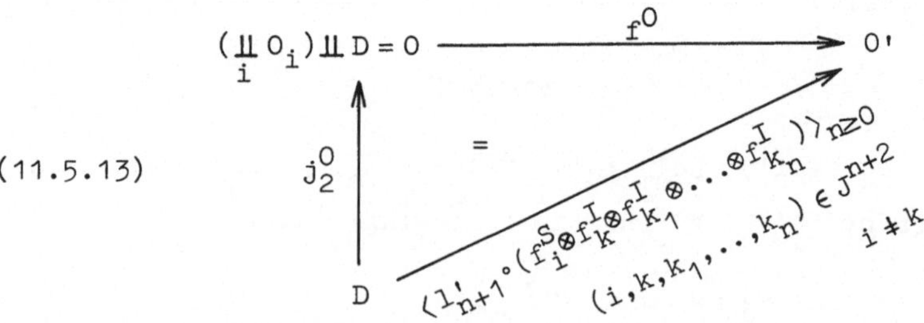

(11.5.13)

where the family $[1'_n]_{n \geq 1}$ is defined by $1'_{n+1} = 1' \cdot (d'_n \otimes I')$.
Hence (11.5.7), (11.5.8), (11.5.9), (11.5.12) and (11.5.13)
together imply the uniqueness of $f = (f^I, f^O, f^S)$.

By the same methods it can be shown that $f = (f^I, f^O, f^S)$ defined by (11.5.7), (11.5.8), (11.5.9), (11.5.12) and
(11.5.13) is in fact an automata morphism which completes
the proof that $A = (I, O, S, d, 1)$ is a coproduct of $[A_i]_{i \in J}$
in $\underline{K}\text{-}\underline{Aut}_V$.

Thus it remains to prove the commutativity of (11.5.10) by
induction on n.
For $n = 0$ the proof can be given using (11.5.6) and a diagram similar to (11.5.14).
Now let be $n \geq 1$ and consider diagram (11.5.14), where the
subdiagram (3) commutes because of $-\amalg D$ being a functor,
(4) by assumption that f is an automata morphism, and the
commutativity of (5) follows from the commutativity of
(11.5.10) for $n := n-1 \geq 0$ and by (11.5.7) in the last component.
This completes the induction step. ∎

For the case $\underline{K} = \underline{Set}$ this proof is already given in [79].

<u>Interpretation</u>: The theorem above is applicable for example
to deterministic, partial, nondeterministic, stochastic, relational, bilinear, topological, and relational topological
automata, cf. 4.2 for the coproducts in the corresponding
categories $\underline{Set}$, $\underline{PD}$, $\underline{ND}$, $\underline{Stoch}$ etc.

For the category $\underline{Set}\text{-}\underline{Aut}$ part (i) of the theorem gives us
a coproduct automaton which is known as the sum composition
of automata (cf. [54]). The state transition and output
function of this automaton work in a single component only,
but we can change this component.
The automaton constructed in part (ii) of the theorem works
in another way: If an input symbol and a state are in the
same A_i, then the state transition and output function d
and 1 (cf. (11.5.3)) are defined as being d_i and 1_i re-

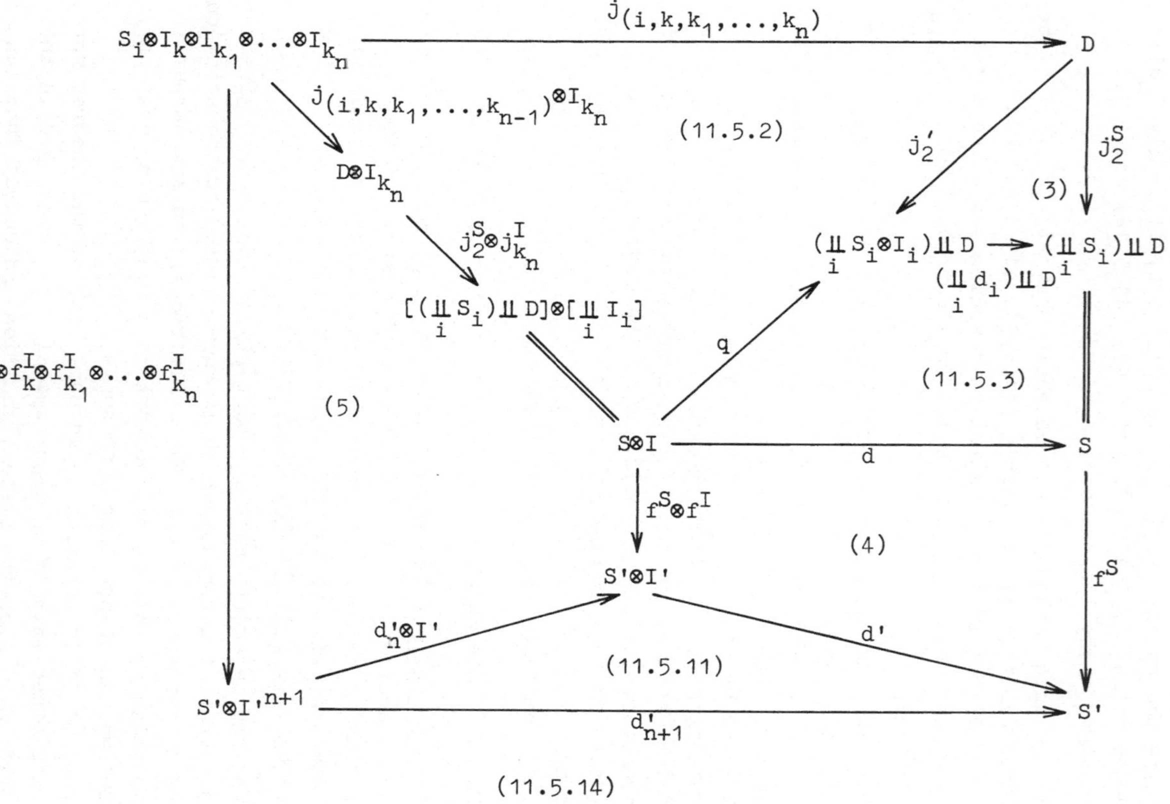

$S_i \otimes I_k \otimes I_{k_1} \otimes \ldots \otimes I_{k_n}$
$\dot{J}(i,k,k_1,\ldots,k_n)$
D
$\dot{J}(i,k,k_1,\ldots,k_{n-1})^{\otimes I_{k_n}}$
(11.5.2)
j_2'
j_2^S
$D \otimes I_{k_n}$
(3)
$j_2^S \otimes j_{k_n}^I$
$(\amalg_i S_i \otimes I_i) \amalg D \longrightarrow (\amalg_i S_i) \amalg D$
$(\amalg_i d_i) \amalg D^i$
$[(\amalg_i S_i) \amalg D] \otimes [\amalg_i I_i]$
q
$f_i^S \otimes f_k^I \otimes f_{k_1}^I \otimes \ldots \otimes f_{k_n}^I$
(11.5.3)
(5)
$S \otimes I$
d
S
$f^S \otimes f^I$
(4)
f^S
$S' \otimes I'$
$d_n' \otimes I'$
d'
(11.5.11)
$S' \otimes I'^{,n+1}$
S'
d_{n+1}'
(11.5.14)

spectively, otherwise d and l work as the identity (up
to changing the coproduct components by an isomorphism).

<u>11.6 Partial and Compactly Generated Automata</u>:
1. There are several other types of automata for which the
structure theory in 11.2 - 11.5 is applicable, for example
partial automata. The category <u>PD</u> of partial functions is
isomorphic to the category <u>Set*</u> of pointed sets as objects
with point-preserving functions as morphisms.
Define the functor $\Phi:\underline{PD} \rightarrow \underline{Set}*$ as being

$\quad$ $\Phi(M) := (M \mathbin{\mathaccent"0\dot{U}} \{*\}, *)$

$\quad$ $\Phi(f:M \rightarrow M') := (\hat{f}:(M \mathbin{\dot{U}} \{*\}, *) \rightarrow (M' \mathbin{\dot{U}} \{*\}, *))$ $\qquad$ with

$\quad$ $\hat{f}(m) = \begin{cases} f(m) & \text{if } f(m) \text{ is defined} \\ * & \text{otherwise} \end{cases}$

and the inverse functor $\Phi^{-1}:\underline{Set}* \rightarrow \underline{PD}$

$\quad$ $\Phi^{-1}(M,*) := M - \{*\}$

$\quad$ $\Phi^{-1}(f:(M,*) \rightarrow (M',*)) := (\hat{f}:(M - \{*\}) \rightarrow (M' - \{*\}))$ $\quad$ with

$\quad$ $\hat{f}(m) = \begin{cases} f(m) & \text{if } f(m) \neq * \\ \text{undefined} & \text{if } f(m) = * \end{cases}$

Similarly as for <u>Set</u> it can be proved that <u>Set*</u> and hence
<u>PD</u> has arbitrary equalizers, products, coequalizers and co-
products. Moreover $(\underline{PD}, \times)$ is closed and hence $-\times I$ preserves coequalizers of kernel pairs. So we can apply 11.2 -
11.5 to partial automata.

2. Let us consider now compactly generated automata consisting of compactly generated Hausdorff-spaces I, O, S and
functions $d:S \times I \rightarrow S$, $1:S \times I \rightarrow O$ which are continuous on
all compact subsets of $S \times I$ endowed with the product topol-
ogy. These automata can be regarded as automata in the
monoidal category $(\underline{CG}, \pi)$ of compactly generated Hausdorff-
spaces as objects and continuous functions as morphisms
where $X \pi Y$ denotes the cartesian product endowed with the
Kelleyfication of the topological product (cf. [32]). It is
well-known (cf. [77]) that this category has all structural
properties mentioned above: It is closed, has equalizers,

products, coequalizers, coproducts and $-\pi-$ preserves coequalizers of kernel pairs. So we can apply 11.2 - 11.5 to
compactly generated automata.

Now we will give the construction of free automata in analogy to that of free monoids or free groups.

<u>11.7 Theorem</u> (F r e e   A u t o m a t a) : Let us assume
that $(\underline{K},\otimes)$ has countable coproducts which are preserved by
$-\otimes K$ and $K\otimes-$ for all $K\in\underline{K}$. Then the forgetful functor

$$V:\underline{K}\text{-}\underline{Aut}_V \longrightarrow \underline{K}^3$$
$$A = (I,O,S,d,l) \longmapsto (I,O,S)$$

has a left adjoint $F:\underline{K}^3 \to \underline{K}\text{-}\underline{Aut}_V$.
This means that for each triplet (I,O,S) of $\underline{K}$-objects
there is an automaton $F(I,O,S)$, called <u>free automaton</u>, and
a triplet of $\underline{K}$-morphisms $u(I,O,S):(I,O,S) \to V(F(I,O,S))$
such that for each triplet of $\underline{K}$-morphisms $f:(I,O,S) \to VA''$
for an arbitrary automaton A'' there is a unique automata
morphism $\overline{f}:F(I,O,S) \to A''$ such that diagram (11.7.1) is
commutative.

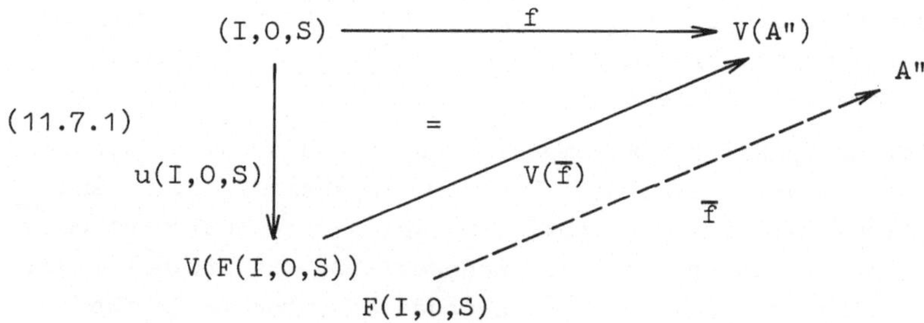

<u>Proof</u>: Define $F(I,O,S):=(I,O',S',d',l')$ by

$$O' := O \amalg (S\otimes I^+) ,$$
$$S' := S\otimes I^* \quad (\text{cf. } (9.3.1)),$$

and diagram (11.7.2) where the isomorphism $I^+ \cong I^*\otimes I$ is

given by (9.3.3), and $u:I*\otimes I \to I*$ by (9.3.2), which can be
interpreted as an inclusion from I^+ into $I*$ and will
also be written in the form $u:I^+ \to I*$.

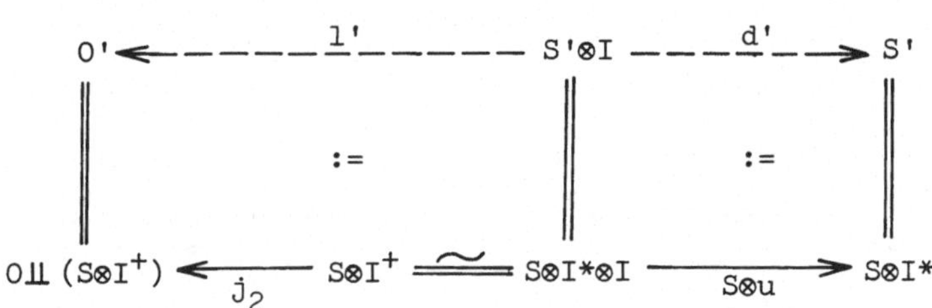

$$(11.7.2)$$

Moreover define $\qquad u(I,0,S) = (u_I, u_0, u_S) =$

$$= (id_I, 0 \xrightarrow{j_1} 0 \amalg (S\otimes I^+) = 0', S = S\otimes U \xrightarrow{S\otimes u_0} S\otimes I*)$$

where $u_0:U \to I*$ is the inclusion (cf. (9.3.1)).

In order to prove the universal property of $F(I,0,S)$ we
take an automaton $A'' = (I'',0'',S'',d'',l'')$ and a triplet
$f = (f_I, f_0, f_S):(I,0,S) \to (I'',0'',S'')$ of $\underline{K}$-morphisms.
We will show first that there is at most one automata mor-
phism $\bar{f} = (\bar{f}_I, \bar{f}_0, \bar{f}_S):F(I,0,S) \to A''$ making (11.7.1) commutative.

Given such a morphism $\bar{f}$ we have: $\bar{f}_I \circ u_I = f_I$, $\bar{f}_0 \circ u_0 = f_0$
and $\bar{f}_S \circ u_S = f_S$ leading to the following equations
(11.7.3), (11.7.4) and (11.7.5):

$$(11.7.3) \qquad\qquad \bar{f}_I = f_I:I \to I''$$

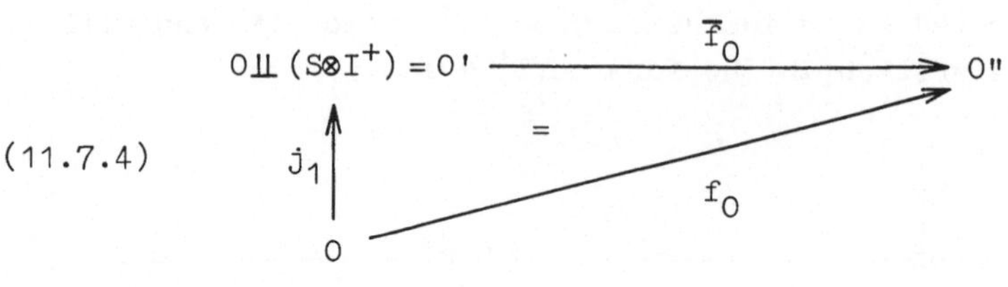

(11.7.4)

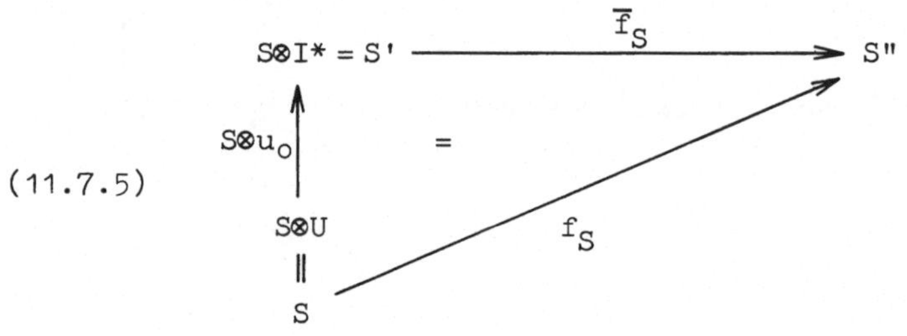

(11.7.5)

So we have to consider

$$S \otimes I^+ \xrightarrow{\;\;j_2\;\;} 0 \amalg (S \otimes I^+) = 0' \xrightarrow{\;\;\overline{f}_0\;\;} 0'' \qquad \text{and}$$

$$S \otimes I^+ \xrightarrow{\;\;S \otimes u\;\;} S \otimes I^* = S' \xrightarrow{\;\;\overline{f}_S\;\;} S'' \; .$$

Now $S \otimes I^* \cong \coprod\limits_{n \geq 0} (S \otimes I^n)$ is a coproduct with injections $S \otimes u_n : S \otimes I^n \to S \otimes I^*$, and the diagram (11.7.6) which will be proved in (11.7.9) to commute for each $n \geq 1$,

(11.7.6)

$$
\begin{array}{ccc}
S \otimes I^n & \xrightarrow{\;\;S \otimes u_n\;\;} & S \otimes I^* = S' \\
{\scriptstyle f_S \otimes f_I^n} \downarrow & = & \downarrow {\scriptstyle \overline{f}_S} \\
S'' \otimes I''^n & \xrightarrow[\;\;d_n''\;\;]{} & S''
\end{array}
$$

together with (11.7.5) yields that (11.7.7) is commutative
and hence $\bar{f}_S$ is uniquely determined by f_I and f_S

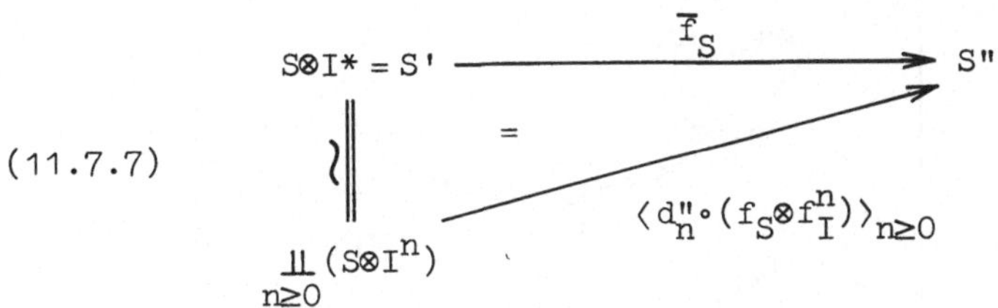

$$(11.7.7)$$

where $d_0'' : S'' \otimes I''^0 = S'' \otimes U \to S''$ is the identity (cf. 1.9) and
d_n'' for $n \geq 1$ is defined by (11.5.11).

Similarly it can be shown that diagram (11.7.8) commutes

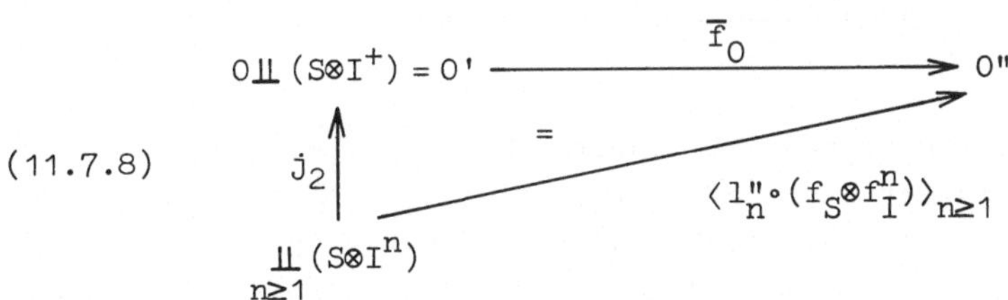

$$(11.7.8)$$

which implies that $\bar{f}_0$ is uniquely determined by f_I , f_0
and f_S .

Hence $\bar{f} = (\bar{f}_I, \bar{f}_0, \bar{f}_S)$ is unique.

By the same methods it can be shown that $\bar{f} = (\bar{f}_I, \bar{f}_0, \bar{f}_S)$
defined by (11.7.3), (11.7.4), (11.7.7) and (11.7.8) is an
automata morphism which makes diagram (11.7.1) commutative.

It remains to show that diagram (11.7.6) is commutative
using induction on n. For $n = 1$ the proof can be given
using (11.7.3) and (11.7.5) and a diagram similar to
(11.7.9).

Now let be $n \geq 2$ and consider (11.7.9).

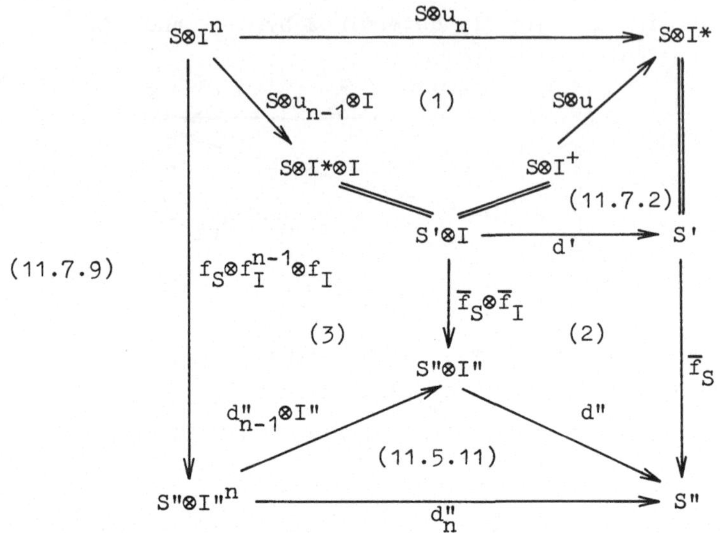

Here the subdiagram (1) commutes by compatibility of the co-
product injections, (2) by the assumption that $\bar{f} = (\bar{f}_I, \bar{f}_0, \bar{f}_S)$
is a morphism of automata and (3) commutes by (11.7.3) and
diagram (11.7.6) for $n-1 \geq 1$. This completes the induction
step. ∎

<u>Remark</u>: The theorem is applicable to deterministic, nonde-
terministic, relational, partial, bilinear, topological and
relational topological automata.
Note that there is no free construction for automata with
fixed I and 0 , i.e. a left adjoint for the forgetful
functor $U: \underline{K\text{-}Aut} \to \underline{K}$, because otherwise U had to preserve
products (cf. 12.9) which is false in general (cf.
11.3 (ii)).

<u>11.8. Characterization of Iso-, Mono- and Epimorphisms of</u>
<u>Automata</u>:
<u>Proposition</u>: (I s o m o r p h i s m s) : The isomorphisms
in $\underline{K\text{-}Aut}$ and $\underline{K\text{-}Aut}_V$ are exactly automata morphisms which are

$\underline{K}$-isomorphisms in each component.

<u>Proof</u>: Let $f = (f_I, f_0, f_S): A \to A'$ be an isomorphism of automata. This implies that f_I, f_0 and f_S are $\underline{K}$-isomorphisms because every functor and hence the forgetful functors $V: \underline{K}\text{-}\underline{Aut}_V \to \underline{K}^3$ and $U: \underline{K}\text{-}\underline{Aut} \to \underline{K}$ preserves isomorphisms.

Vice versa, if $f: A \to A'$ is an automata morphism such that f_I, f_0 and f_S are $\underline{K}$-isomorphisms, we have the equation
$$[d \circ (f_S^{-1} \otimes f_I^{-1})] \circ (f_S \otimes f_I) = d = f_S^{-1} \circ f_S \circ d = [f_S^{-1} \circ d'] \circ (f_S \otimes f_I)$$
and hence $d \circ (f_S^{-1} \otimes f_I^{-1}) = f_S^{-1} \circ d'$ and similarly
$$l \circ (f_S^{-1} \otimes f_I^{-1}) = f_0^{-1} \circ l' .$$
This implies that $f^{-1} := (f_I^{-1}, f_0^{-1}, f_S^{-1})$ is an automata morphism and hence f is an isomorphism of automata.
Note that the same proof is valid for automata with fixed I and O.   ∎

<u>Corollary</u> (M o n o m o r p h i s m s) : Under the same assumptions as in theorem 11.7 the monomorphisms in $\underline{K}\text{-}\underline{Aut}_V$ are exactly automata morphisms which are monomorphisms in each component.

<u>Proof</u>: By the theorem 11.7 and 12.9 $V: \underline{K}\text{-}\underline{Aut}_V \to \underline{K}^3$ preserves monomorphisms.
Vice versa an automata morphism which is a monomorphism in each component is clearly a monomorphism in $\underline{K}\text{-}\underline{Aut}_V$.   ∎

<u>Remark</u> (R e g u l a r   E p i m o r p h i s m s) :
It seems to be more difficult to characterize all epimorphisms in categories of automata. But we have a characterization for regular epimorphisms. An epimorphism is called <u>regular</u> if it is the coequalizer of some pair, and we have the following categorical theorem which is proved in [65] in a more general context:

<u>Theorem</u>: If $(\underline{K}, \otimes)$ has kernel pairs and coequalizers of kernel pairs and $-\otimes-$ preserves coequalizers of kernel pairs

then the regular epimorphisms in $\underline{K}\text{-}\underline{Aut}_V$ are exactly automata morphisms which are regular epimorphisms in each component.

We can apply this theorem for example to deterministic, partial and compactly generated topological automata (for the last type cf. [32]). Note that in $\underline{K} = \underline{Set}$ each epimorphism is regular.

11.9 <u>Factorization of Automata Morphisms</u>: We assume that $(\underline{K},\otimes)$ has an $\mathfrak{E}\text{-}\mathfrak{M}$-factorization which satisfies $K\otimes e\in\mathfrak{E}$ and $e\otimes K\in\mathfrak{E}$ for all $K\in\underline{K}$, $e\in\mathfrak{E}$ (cf. 4.6).

<u>Theorem</u> (L i f t i n g   o f   t h e $\mathfrak{E}$ - $\mathfrak{M}$ - F a c - t o r i z a t i o n) : $\underline{K}\text{-}\underline{Aut}$ and $\underline{K}\text{-}\underline{Aut}_V$ have $\mathfrak{E}\text{-}\mathfrak{M}$-factorizations which can be constructed componentwise.

<u>Proof</u>: The proof will be given only for $\underline{K}\text{-}\underline{Aut}$. It is similar for $\underline{K}\text{-}\underline{Aut}_V$. Let $f:A \to A'$ be an automata morphism in $\underline{K}\text{-}\underline{Aut}$ and $m\circ e:S \to S'' \to S'$ the $\mathfrak{E}\text{-}\mathfrak{M}$-factorization in $\underline{K}$ of $f:S \to S'$. By the diagonal lemma (cf. 4.7,1) and our assumption $e\otimes I\in\mathfrak{E}$ there exist unique morphisms d'' and l'' such that diagram (11.9.1) is commutative. This yields that (S'',d'',l'') is an automaton and e and m are automata morphisms. The uniqueness of the decomposition $m\circ e = f$ in $\underline{K}\text{-}\underline{Aut}$ up to isomorphism follows from the uniqueness in $\underline{K}$ and 11.8.

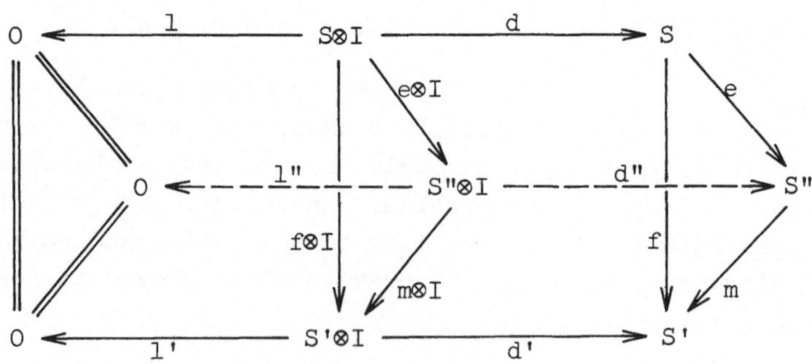

$$(11.9.1)$$

12. Appendix: Basic Notions of Category Theory

In this appendix we summarize all those basic categorical
notions which are used, but only informally introduced, in
this book. For motivation and examples we refer to the previous chapters (cf. subject index). But we do not reformu-
late the definition of monoidal categories and $\mathfrak{E}$-$\mathfrak{M}$-factor-
izations for example because the exact definitions and all
necessary explanations are already given in the corresponding chapters. Moreover we give the proof for the result that
left adjoint functors preserve colimits for the special case
of cointersections which is used in chapter 7 . A more detailed version of categorical notions and results with respect to automata theory is given in [33]. For further categorical investigations including the set theoretical founda-
tions we advise the following books: [50,59,63,71].

<u>12.1 Categories</u>: A <u>category</u> $\underline{K}$ consists of a class $|\underline{K}|$ of
<u>objects</u>, for each pair $A,B\in|\underline{K}|$ a set $Mor\underline{K}(A,B)$ of
<u>morphisms</u>, written $f:A \to B$ for $f\in Mor\underline{K}(A,B)$, and a
<u>composition</u> $\circ:Mor\underline{K}(A,B)\times Mor\underline{K}(B,C) \to Mor\underline{K}(A,C)$
$$(f:A \to B , g:B \to C) \mapsto (g\circ f:A \to C)$$

(more precisely a family of functions $\circ_{A,B,C}$ for all objects A, B, C) such that the following axioms are satisfied:

(Ass) $(h\circ g)\bullet f = h\circ(g\circ f)$ for all morphisms f, g, h   if
at least one side is defined.

(Id) For each object $A\in|\underline{K}|$ there is a morphism
$id_A\in Mor\underline{K}(A,A)$, called <u>identity</u> of A , such that we
have for all $f:A \to B$ and $g:C \to A$ with $B,C\in|\underline{K}|$:
$f\circ id_A = f$ and $id_A\circ g = g$.

<u>Conventions</u>: Objects in $\underline{K}$ will be denoted by capital let-
ters and morphisms by small ones. Very often we will write
A∈$\underline{K}$ and f∈$\underline{K}$ for an object A and a morphism f in $\underline{K}$
respectively. For a morphism f:A → B   A is called <u>domain</u>
and B <u>codomain</u> of f .

A <u>subcategory</u> $\underline{K}'$ of a category $\underline{K}$ consists of a subclass
|$\underline{K}'$| of |$\underline{K}$| as objects and subsets Mor$\underline{K}'$(A,B) of
Mor$\underline{K}$(A,B) as morphisms in $\underline{K}'$ for all A,B∈|$\underline{K}'$|. The com-
position in the category $\underline{K}'$ is the restriction of the
composition in $\underline{K}$.
A subcategory $\underline{K}'$ of $\underline{K}$ is called <u>full</u> if we have:
Mor$\underline{K}'$(A,B) = Mor$\underline{K}$(A,B) for all A,B∈$\underline{K}'$.
A category $\underline{K}$ is called <u>discrete</u> if there are no morphisms
in $\underline{K}$ except identities, and it is called <u>partially ordered</u>
if for all objects A,B∈$\underline{K}$ there is at most one $\underline{K}$-morphism
f:A → B and the existence of morphisms f:A → B and g:B → A
implies A = B . For f:A → B we will write A ⊆ B and "⊆"
defines a partial order of the object class |$\underline{K}$|. Vice versa
each partially ordered class ($\underline{B}$,≤) can be regarded as a
partially ordered category where Mor$\underline{B}$(x,y) for x,y∈$\underline{B}$ consists of one element in the case x≤ y and is empty otherwise.

<u>12.2 Diagrams, Duality</u>: Given morphisms f:A → X , g:X → B ,
f':A → Y , g':Y → B the equality g∘f = g'∘f' can be
illustrated by the following commutative diagram.

(12.2.1)

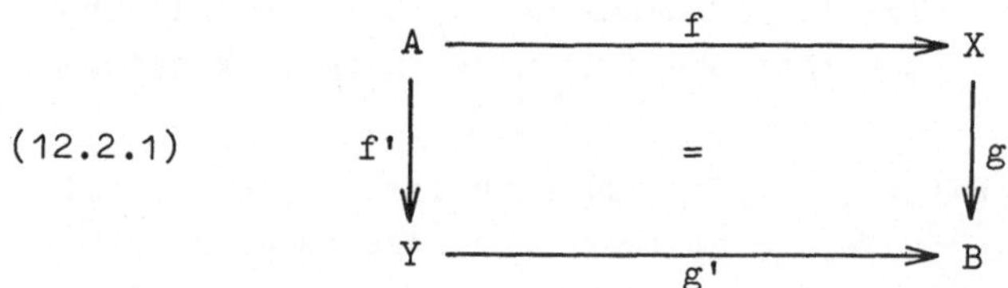

The sign "=" means that the diagram is commutative.
Reversing all arrows in a diagram we get the <u>dual diagram</u>.

Roughly speaking this procedure allows to "<u>dualize</u>" all categorical notions and constructions, because they are defined by commutative diagrams in general.

Dualizing diagram (12.2.1) we get:

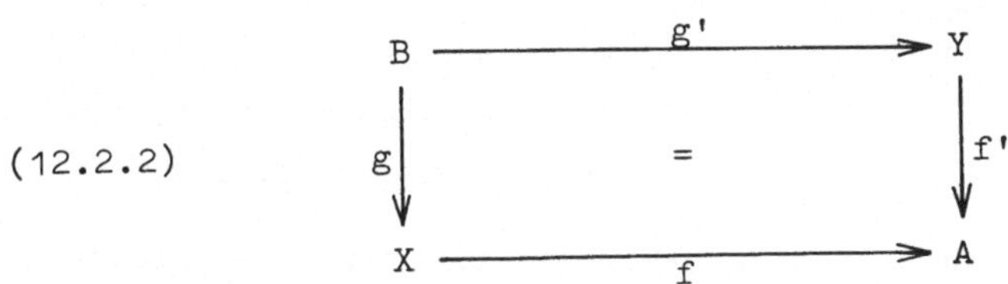

(12.2.2)

Thus the "dual statement" of $g \circ f = g' \circ f'$ for example is $f \circ g = f' \circ g'$.

Although we do not state the duality principle in detail dual notions and constructions will be used frequently.

<u>12.3 Iso-, Mono-, Epimorphisms</u>: A morphism $f : A \to B$ is called <u>isomorphism</u> if there exists a morphism $g : B \to A$ with $f \circ g = id_B$ and $g \circ f = id_A$. Two objects are called <u>isomorphic</u>, written $A \cong B$, if there is an isomorphism $f : A \to B$. If we only have $f \circ g = id_B$ or $g \circ f = id_A$ f is called <u>retraction</u> or <u>coretraction</u> respectively.

A morphism $f : A \to B$ is called <u>monomorphism</u> if for all $g, h : C \to A$ satisfying $f \circ g = f \circ h$ we have $g = h$ (left cancellation). Dually $f : B \to A$ is called <u>epimorphism</u> if for all $g, h : A \to C$ with $g \circ f = h \circ f$ we have $g = h$ (right cancellation).

The dual notion of retraction is coretraction and vice versa, that of an isomorphism is dual to itself.

<u>Remark</u>: 1. The above defined notions imply each other in the following way:

isomorphism
→ coretraction ⟶ monomorphism
→ retraction ⟶ epimorphism

2. Given two mono- (epi-) morphisms $f:A \to B$, $g:B \to C$, $g \circ f$
is a mono- (epi-) morphism . Vice versa,if $g \circ f$ is a monomorphism so is f. Dually g is an epimorphism if $g \circ f$ has
this property.

<u>12.4 Products, Coproducts</u>: Let $[A_i]_{i \in I}$ be a family of ob-
jects in <u>K</u>. An object $A \in \underline{K}$ together with a family
$[p_i:A \to A_i]_{i \in I}$ of morphisms is called <u>product</u> of $[A_i]_{i \in I}$,
written $A = \underset{i \in I}{\Pi} A_i$, if for each $K \in \underline{K}$ and each family

$[f_i:K \to A_i]_{i \in I}$ of morphisms there is exactly one morphism
$f:K \to A$ satisfying $p_i \cdot f = f_i$ for all $i \in I$. The morphisms p_i
are called <u>projections</u>.

Dually $(A,[u_i:A_i \to A]_{i \in I})$ is called <u>coproduct</u> of the fami-
ly $[A_i]_{i \in I}$ (written $A = \underset{i \in I}{\amalg} A_i$) with <u>injections</u> u_i if for

each $K \in \underline{K}$ and each family $[f_i:A_i \to K]_{i \in I}$ there is exactly
one morphism $f:A \to K$ in (12.4.1) with $f \circ u_i = f_i$ for all
$i \in I$.

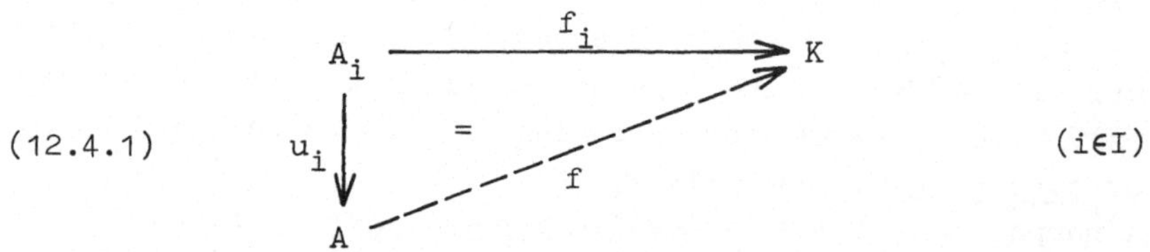

The uniqueness of f in (12.4.1) implies the following important concept to prove equality of morphisms:

<u>Corollary</u>: Two morphisms $f,f':A \to K$ are equal iff we have
$f \circ u_i = f' \circ u_i$ for all $i \in I$.

If the set I is empty the coproduct object A is called
<u>initial</u> object meaning that for each object K in <u>K</u> there
is exactly one morphism f from A to K .

Dually the corresponding product object A' is called
__final__ or __terminal__ object meaning that there is exactly one
morphism from each $\underline{K}$-object K to A' .

__12.5 Functors__: Let $\underline{K}$ and $\underline{L}$ be categories. $F:\underline{K} \to \underline{L}$ is
called a __functor__ if F assigns to each object A in $\underline{K}$
exactly one object F(A) in $\underline{L}$ and to each morphism $f:A \to B$
in $\underline{K}$ an $\underline{L}$-morphism $F(f):F(A) \to F(B)$ such that the following axioms are satisfied:

F1) $F(g \circ f) = F(g) \circ F(f)$ for all $g \circ f$ in $\underline{K}$.

F2) $F(id_A) = id_{F(A)}$ for all objects A in $\underline{K}$.

The __inclusion__ of a subcategory $\underline{K}'$ of $\underline{K}$ is a functor
which is called __identity functor__ $Id_{\underline{K}}$ in the case $\underline{K}' = \underline{K}$.

Replacing $\underline{K}$ by $\underline{K} \times \underline{K}$ we get the notion of a __bifunctor__
$F:\underline{K} \times \underline{K} \to \underline{L}$ assigning to each pair (A_1,A_2) of objects in $\underline{K}$
an object $F(A_1,A_2)$ in $\underline{L}$ and to each pair of morphisms
$(f_1:A_1 \to B_1, f_2:A_2 \to B_2)$ in $\underline{K}$ an $\underline{L}$-morphism
$F(f_1,f_2):F(A_1,A_2) \to F(B_1,B_2)$ such that we have

B1)

$$F(A_1,A_2) \xrightarrow{\ F(f_1,f_2)\ } F(B_1,B_2) \xrightarrow{\ F(g_1,g_2)\ } F(C_1,C_2)$$
$$\underbrace{\qquad\qquad\qquad\qquad\qquad\qquad\qquad}_{F(g_1 \circ f_1, g_2 \circ f_2)} \quad = \quad \Big\uparrow \ .$$

for all $g_1 \circ f_1$ and $g_2 \circ f_2$ in $\underline{K}$ and

B2) $F(id_{A_1}, id_{A_2}) = id_{F(A_1,A_2)}$ for all A_1, A_2 in $\underline{K}$.

For each bifunctor $F:\underline{K} \times \underline{K} \to \underline{L}$ we have partial functors
$F(A_1,-):\underline{K} \to \underline{L}$ and $F(-,A_2):\underline{K} \to \underline{L}$ for all A_1, A_2 in $\underline{K}$
defined in the obvious way, e.g.
$F(A_1,-)(A_2) = F(-,A_2)(A_1) = F(A_1,A_2)$.

The composition of functors $F:\underline{K} \to \underline{L}$ and $G:\underline{L} \to \underline{M}$ is defined by $G \circ F(A) = G(F(A))$ and $G \circ F(f) = G(F(f))$ for objects

and morphisms respectively leading to the composite functor
$G \circ F : \underline{K} \to \underline{M}$.

<u>Convention</u>: Sometimes we will drop some brackets writing
$GF(A)$ instead of $G(F(A))$ or FA instead of $F(A)$ for
example.

<u>12.6 Isomorphisms and Natural Transformations</u>: Two categories $\underline{K}$ and $\underline{L}$ are called <u>isomorphic</u> if there are two
functors $F : \underline{K} \to \underline{L}$ and $G : \underline{L} \to \underline{K}$ such that $G \bullet F = Id_{\underline{K}}$ and
$F \circ G = Id_{\underline{L}}$.

Let $F : \underline{K} \to \underline{L}$ and $G : \underline{K} \to \underline{L}$ be functors. Then a family
$u = [u(K) : F(K) \to G(K)]_{K \in \underline{K}}$ of $\underline{L}$-morphisms is called <u>natural
transformation</u> $u : F \to G$ if for all $\underline{K}$-morphisms $f : K \to K'$
the diagram (12.6.1) is commutative.

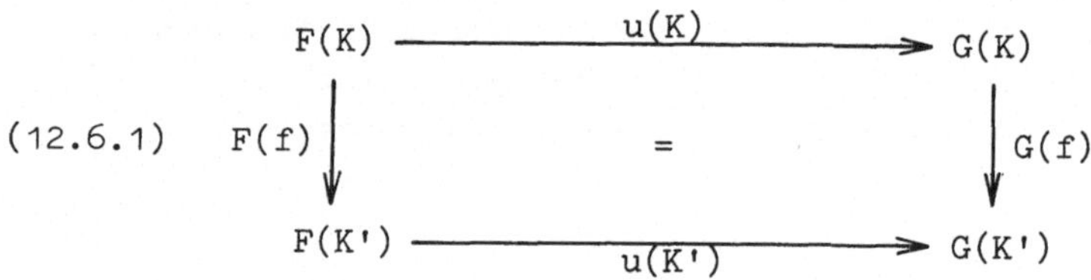

u is called <u>natural isomorphism</u> if all $u(K)$ are $\underline{L}$-isomorphisms. Sometimes the components $u(K)$ are also called natural isomorphisms. In this case F and G are called
<u>natural equivalent</u>.

<u>12.7 Adjoint Functors</u>: Let $T : \underline{K} \to \underline{G}$ be a functor. The
<u>universal problem</u> for the functor T is called solvable if
for all $G \in \underline{G}$ there is an object $S(G)$ in $\underline{K}$ and a (universal) $\underline{G}$-morphism $u(G) : G \to TS(G)$ such that for all $K \in \underline{K}$
and all $\underline{G}$-morphisms $f : G \to T(K)$ there is exactly one $\underline{K}$-morphism $\bar{f} : S(G) \to K$ such that the diagram (12.7.1) is commutive.

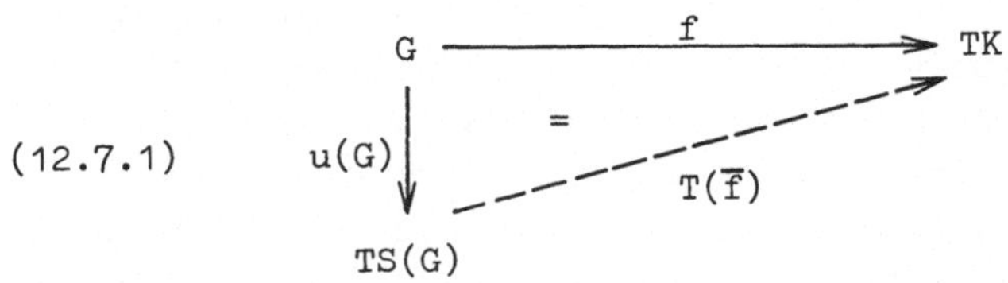

$$(12.7.1)$$

Remark: There is a unique $\underline{K}$-morphism $S(g):S(G) \to S(G')$ for each $\underline{G}$-morphism $g:G \to G'$ such that (12.7.2) is commutative, this means that $S(g)$ is defined by (12.7.1) for $f := u(G') \circ g$.

$$(12.7.2)$$

Thus we get a functor $S:\underline{G} \to \underline{K}$ and a natural transformation $u:\mathrm{Id}_{\underline{G}} \to T \circ S$ using the uniqueness of $S(G)$ in (12.7.2). This functor S is called left adjoint to T, written $S \dashv T$, and u is called the unit of the adjunction $S \dashv T$.

Dually, starting with a functor $S:\underline{G} \to \underline{K}$ the couniversal problem for S is called solvable if for each $K \in \underline{K}$ there exists an object $T(K)$ in $\underline{G}$ and a (couniversal) $\underline{K}$-morphism $c(K):ST(K) \to K$ such that for all $G \in \underline{G}$ and $\bar{f}:S(G) \to K$ in $\underline{K}$ there is exactly one $f:G \to T(K)$ satisfying $c(K) \circ S(f) = \bar{f}$. In analogue to the universal problem T can be extended to a functor $T:\underline{K} \to \underline{G}$ such that c becomes a natural transformation $c:S \circ T \to \mathrm{Id}_{\underline{K}}$. Now T is called right adjoint to S, written $S \dashv T$, and c is called counit of the adjunction S⟶T.

<u>Remarks</u>: 1. The functor S solving the universal problem
for T is uniquely determined up to natural equivalence.
Vice versa given S , the solution T of the couniversal
problem is also unique up to natural equivalence.
2. The universal problem for T is solvable with left adjoint functor S iff the couniversal problem for S is
solvable with the right adjoint functor T.

<u>Composition of Adjoint Functors</u>: Let $T:\underline{K} \to \underline{G}$ and $T':\underline{L} \to \underline{K}$
be functors with left adjoint functors $S:\underline{G} \to \underline{K}$ and
$S':\underline{K} \to \underline{L}$ and units u and u' respectively. Then
$S' \circ S:\underline{G} \to \underline{L}$ is left adjoint to $T \circ T':\underline{L} \to \underline{G}$ with unit
$u'':Id_{\underline{G}} \to T \circ T' \circ S' \circ S$ defined by $u''(G) := T(u'(S(G))) \circ u(G)$.

<u>12.8 Comma Categories</u>: Let $\underline{K}$ be a category and K_o a
fixed object in $\underline{K}$. Then the <u>comma category</u> $(\underline{K} \downarrow K_o)$ is defined as follows:
Objects of $(\underline{K} \downarrow K_o)$ are the $\underline{K}$-morphisms $f:K \to K_o$ with arbitrary K in $\underline{K}$. Morphisms in $(\underline{K} \downarrow K_o)$ from $f:K \to K_o$ to
$f':K' \to K_o$ are $\underline{K}$-morphisms $g:K \to K'$ with $f' \circ g = f$. The
composition of morphisms g and g' in $(\underline{K} \downarrow K_o)$ is defined
by the composition $g' \bullet g$ in $\underline{K}$ which is again a morphism in
$(\underline{K} \downarrow K_o)$ by

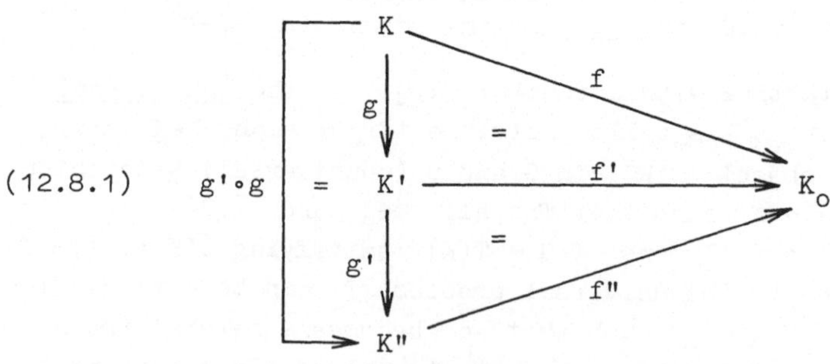

<u>12.9 Special Limits</u>: A morphism $d:D \to A$ is called <u>equal-
izer</u> of the morphisms $f,g:A \to B$ if $f \circ d = g \circ d$ and for all
morphisms $h:X \to A$ with codomain A and $f \circ h = g \circ h$ there
is exactly one morphism $\bar{h}:X \to D$ in (12.9.1) such that
$d \circ \bar{h} = h$.

(12.9.1)

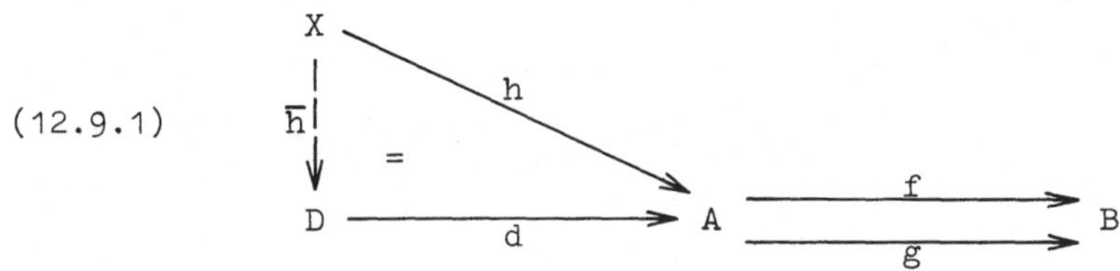

Dually a morphism $d:A \to D$ is called <u>coequalizer</u> of the
morphisms $f,g:B \to A$ if $d \circ f = d \circ g$ and for all morphisms
$h:A \to X$ with $h \circ f = h \circ g$ there is exactly one morphism
$\bar{h}:D \to X$ satisfying $\bar{h} \circ d = h$.

Given a set I and a family $[f_i:A_i \to B]_{i \in I}$ of morphisms
an object P together with a family $[p_i:P \to A_i]_{i \in I}$ is
called <u>fibre product</u> of $[f_i]_{i \in I}$ if the following conditions are satisfied:

(i) $f_i \circ p_i = f_j \circ p_j$ for all $i,j \in I$.
(ii) For each family $[h_i:X \to A_i]_{i \in I}$ satisfying (i) there
     is exactly one morphism $h:X \to P$ with $p_i \circ h = h_i$ for
     all $i \in I$ in (12.9.2)

If I is an arbitrary class and not necessary a set in the
Neumann-Gödel-Bernays set theory and all morphisms f_i $(i \in I)$
are monomorphisms then the fibre product is called <u>large
intersection</u>.

In the case $I = \{1,2\}$ the fibre product is called <u>pullback</u>
of f_1 and f_2 or <u>kernel pair</u> of f if we have $A_1 = A_2 = A$
and $f_1 = f_2 = f$ in addition. If f_1 is a monomorphism the
pullback is called <u>inverse image</u> of f_1 under f_2 .

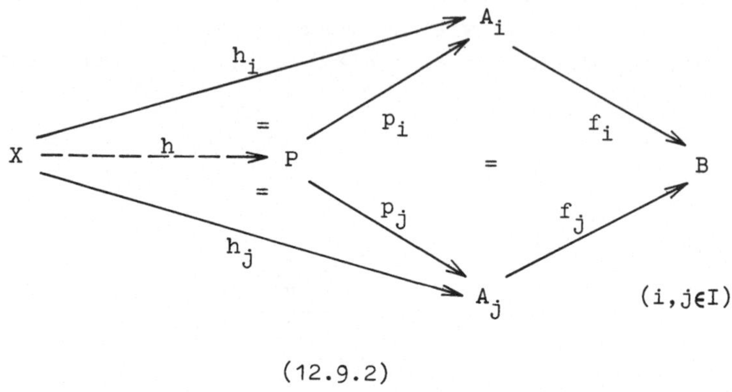

(12.9.2)

Reversing all arrows we get the dual notions (cf. 12.2) of
fibre product, pullback and kernel pair called <u>cofibre prod-
uct</u>, <u>pushout</u> and <u>cokernel pair</u>. Dually to large intersections a <u>large cointersection</u> is a cofibre product of a family $[f_i : B \to A_i]_{i \in I}$ of epimorphisms where I is an arbitrary class.

<u>Remarks</u>: 1. Products, equalizers , fibre products, pullbacks,
kernel pairs, inverse images and (large) intersections are
special cases of a more general notion called <u>limit of a
(large) diagram</u> which is not necessary to introduce here
(cf. [33,50,59,63,71]). The dual notions can be generalized
to a <u>colimit of a (large) diagram</u>.
2. Limits and colimits of diagrams are unique up to isomorphism and they exist iff we have the existence of products
and equalizers or coproducts and coequalizers respectively.
Especially the pullback of $f_1 : A_1 \to B$ and $f_2 : A_2 \to B$ can
be constructed in the following way:
Take the product $A_1 \pi A_2$ of A_1 and A_2 with the projections $p_1 : A_1 \pi A_2 \to A_1$ and $p_2 : A_1 \pi A_2 \to A_2$ and construct
the equalizer $d : P \to A_1 \pi A_2$ of the pair
$f_1 \circ p_1, f_2 \circ p_2 : A_1 \pi A_2 \to B$. Then P together with $q_1 := p_1 \circ d$

225

and $q_2 := p_2 \circ d$ is the pullback of f_1 and f_2 in (12.9.3).

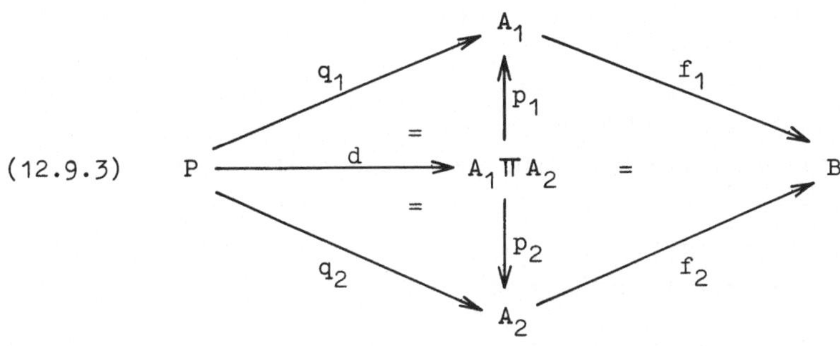

(12.9.3)

3. Let S be left adjoint to T (cf. 12.7) then S preserves arbitrary colimits, especially coproducts and coequalizers, and T preserves all limits dually (cf. [33,50,59, 63,71]). For the special case of cointersections which is needed in 7.3 this will be proved by the following lemma:

<u>12.10 Lemma</u>: Let $S:\underline{G} \to \underline{K}$ be the left adjoint functor of $T:\underline{K} \to \underline{G}$ then S preserves cofibre products, i.e. given a set I and a family $[f_i:G \to G_i]_{i \in I}$ of $\underline{G}$-morphisms which has a cofibre product $\bar{G}$ with $[u_i:G_i \to \bar{G}]_{i \in I}$ in $\underline{G}$ the object $S(\bar{G})$ together with $[S(u_i):S(G_i) \to S(\bar{G})]_{i \in I}$ is a cofibre product of the family $[S(f_i):S(G) \to S(G_i)]_{i \in I}$ in $\underline{K}$.

<u>Corollary</u>: S preserves (large) cointersections of arbitrary families $[f_i:G \to G_i]_{i \in I}$ of $\underline{G}$-epimorphisms for each set (class) I.

<u>Proof</u>: In order to verify that $S(\bar{G})$ with $[S(u_i):S(G_i) \to S(\bar{G})]_{i \in I}$ is the cofibre product of $[S(f_i):S(G) \to S(G_i)]_{i \in I}$ we have to verify dually to (12.9.2)

a) $S(u_i) \circ S(f_i) = S(u_j) \circ S(f_j)$ for all $i,j \in I$,

b) for each family $[k_i:S(G_i) \to K]_{i \in I}$ satisfying a) there is exactly one morphism $k:S(\bar{G}) \to K$ with $k \bullet S(u_i) = k_i$.

By assumption we have $u_i \circ f_i = u_j \circ f_j$ for all $i,j \in I$ because $\bar{G}$ together with $[u_i]_{i \in I}$ is the cofibre product of $[f_i]_{i \in I}$. Thus we have a) using the functor property F1 in 12.5 for S. Now given a family $[k_i : S(G_i) \to K]_{i \in I}$ with $k_i \circ S(f_i) = = k_j \circ S(f_j)$ $(i,j \in I)$ we define for each $i \in I$ $g_i := T(k_i) \circ u(G_i)$ where $u(G_i) : G_i \to TS(G_i)$ is the universal morphism of G_i (cf. (1) in (12.10.1)). Thus we have for all $i,j \in I$ by (12.7.2)

$$g_i \circ f_i = T(k_i) \circ u(G_i) \circ f_i = T(k_i) \circ TS(f_i) \bullet u(G) = T(k_i \circ S(f_i)) \circ u(G)$$
$$= T(k_j \circ S(f_j)) \circ u(G) = g_j \bullet f_j \ .$$

Hence there is a unique $g : \bar{G} \to TK$ such that (2) in (12.10.1) commutes for all $i \in I$ using the universal property of the cofibre product $\bar{G}$. By (12.7.1) we get a unique $k : S(\bar{G}) \to K$ such that (3) in (12.10.1) is commutative. We want to show that k satisfies condition b) above. Using the commutativity of (12.10.1) except (4) we get $T(k_i) \circ u(G_i) = = T(k) \circ TS(u_i) \circ u(G_i)$ for all $i \in I$ which implies $k_i = k \circ S(u_i)$ using the uniqueness properties of $u(G_i)$ for all $i \in I$ (cf. (12.7.1)). It remains to show that for each $k' : S(\bar{G}) \to K$ with $k' \circ S(u_i) = k_i$ for all $i \in I$ we have $k' = k$. Thus re-

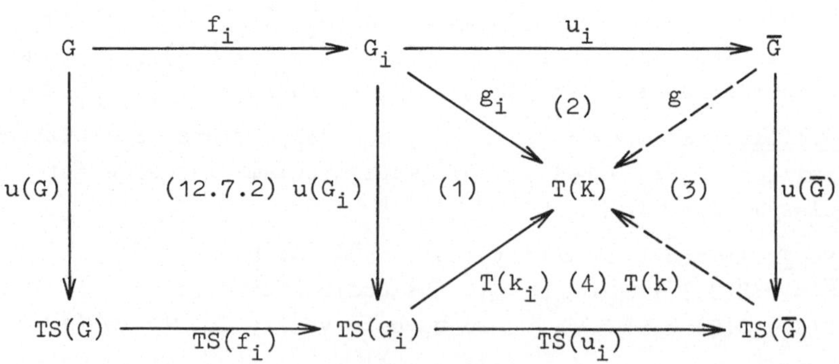

(12.10.1)

placing k by k' in (12.10.1) we have commutativity in
(12.10.1) except (3) . But (3) is also commutative and hence
k' = k by (12.7.1) using the uniqueness property of the cofibre product $\bar{G}$ with the family $[u_i]_{i\in I}$ of "universal
morphisms". Thus we have shown that S preserves cofibre
products and hence cointersections assuming that all
$f_i:G \to G_i$ (i∈I) are $\underline{G}$-epimorphisms because it is easy to
show that $S(f_i):S(G) \to S(G_i)$ (i∈I) are $\underline{K}$-epimorphisms in
this case. But because we do not need this property in 7.3
we do not verify it explicitly. Finally let us remark that
the set I can be replaced by a proper class in our proof
such that S preserves also large cofibre products and
large cointersections if they are existing in $\underline{G}$.   ∎

Special Symbols

General Notions and Abbreviations

[27]	reference 27	i.e.	id est, that means
6.3	section 6.3	e.g.	for example
3.4,1	statement 1 in	iff	if and only if
	section 3.4	cf.	confer
(9.3.7)	diagram (9.3.7) in	resp.	respective(ly)
	section 9.3	■	end of proof

Set Theoretical Symbols

{ }	brackets for sets	$\langle A,B\rangle$	set of all functions
$\in$	element		from A to B
$\times$, $\bigtimes$	cartesian product	$S/_R$	quotient set
$\cup$, $\bigcup$	union	$[x]$	equivalence class of x
$\dot{\cup}$, $\dot{\bigcup}$	disjoint union	$A - B$	difference, complement
$\cap$, $\bigcap$	intersection	$\subseteq$	inclusion
$\emptyset$	empty set	$\leq$	less or equal, partial
$\wp A$	power set of A		order
$\wp'A$	$\wp A$ without the	card	cardinality
	empty subset	Σ	sum
$:=$	defining equation	$\mathbb{N}$	natural numbers
		$\mathbb{N}_o$	$\mathbb{N} \cup \{0\}$

Categorical Symbols

$\underline{K}$, $\underline{L}$, $\underline{B}$	categories 12.1	$(\underline{K},\otimes)$	monoidal category 1.9
$\underline{Set}$	1.10	$(\underline{K},\otimes,U)$	
$\underline{PD}$	1.13	$\otimes$	tensor product 1.9
$\underline{Mod}_R$	1.10	$g \circ f$	composition 12.1
$\underline{Top}$	1.10	id_B	identity 12.1
$\underline{ND}$	1.10	$A \cong B$	isomorphism 12.3,12.6
$\underline{Rel}$	1.13	$A \xrightarrow{\sim} B$	
$\underline{Stoch}$	1.10	f^{-1}	inverse morphism 3.2

$\mathfrak{E}$, $\mathfrak{M}$	class of epi- resp. monomorphisms 4.6	$\subseteq$, $\leq$	partial order 12.1,3.1
		$(\underline{K} \downarrow K_o)$	comma category 12.8
$\tilde{\mathfrak{E}}$	class of surjective morphisms 9.1	$Id_{\underline{K}}$	identity functor 12.5
		$\langle A,B\rangle$	$:= \langle A,-\rangle(B)$ 4.3
Π	product 12.4	ev	evaluation 4.3
$\amalg$	coproduct 12.4	P , J	functors 6.2
$[K_i]_{i \in I}$	family of objects 12.4	v , v_o	couniversal morphism 6.2
$[f_i]_{i \in I}$	family of morphisms 12.4		
		$S \longrightarrow T$	adjoint functors 12.7
$\langle f_i \rangle_{i \in I}$	induced morphism 4.2	$\underline{\underline{S}}$	systematic 3.1

<u>Automata Theoretical Symbols</u>

$A = (I,O,S,d,l)$	automaton	$\underline{B}$	behavior category 3.1
$A = (S,d,l)$	1.1, 1.11	E	behavior functor 3.1
$A = (S,d)$	Medvedev-automaton 1.14	$M^*(b)$	minimal realization 3.6, 5.3, 9.5, 9.10
$A' = (S,d,l,a)$	initial automaton 9.2,9.4	$F(b)$	free realization 9.4
		$S(b)$	states of the behavior 9.6, 9.10
$\breve{P}A$	power automaton 8.2		
$A(m,n)$	scoop automaton 10.3	$\underline{K}\text{-}\underline{Aut}$	category of automata 1.11
$w = x_1 \ldots x_n$	string 1.2		
$\square$	empty string 1.2	$(\underline{K},\otimes)\text{-}\underline{Aut}$	1.11
I^+	free semigroup 1.2,4.2	$\underline{K}\text{-}\underline{K}'\text{-}\underline{Aut}$	1.11
l^+	extended l 1.2,4.2,6.3	$\underline{K}\text{-}\underline{K}'\text{-}\underline{Aut}^w,\ldots$	7.6, 9.9
I^*	free monoid 1.2,5.6, 9.3	$\underline{K}\text{-}\underline{Aut}^{\cdot},\ldots$	9.2, 9.1, 9.5, 9.9, 9.10
a^*	reachability morphism 9.4	$\underline{K}'\text{-}\underline{Aut}(I,PO)$	8.1
		$\underline{K}\text{-}\underline{Aut}_v$	11.1
$M(A)$	machine morphism 2.1, 4.3, 6.3	$\underline{K}\text{-}\underline{Medv}$	1.14
$E(A)$	behavior 2.1,4.6,6.3	$\underline{\underline{K}}\text{-}\underline{\underline{Aut}}$	systematic of automata 5.2, 5.4, 5.7, 7.6, 7.7, 7.8, 9.2, 9.5, 9.8, 9.9
$E^{\cdot}(A)$	initial behavior 9.2		
$R(A)$	reduced automaton 5.5, 7.3		

References

[1] ALAGIC,S., Categorical Theory of Tree Processing,
in: [60], 80 - 88

[2] ARBIB,M.A., Theories of Abstract Automata,
Englewood Cliffs, N.J. 1969

[3] ARBIB,M.A. / GIVEON,Y., Algebra Automata I and II,
Inf. Contr. $\underline{12}$ (1968), 331 - 370

[4] ARBIB,M.A. / MANES,E.G., Machines in a Category: An Ex-
pository Introduction, Techn. Rep. 71 C - 6
COINS, Univ. of Mass. at Amherst (1972),
to appear in SIAM Review

[5] - - Adjoint Machines, State-Behavior, and Duality,
Techn. Rep. 73 B - 1 COINS, Univ. of Mass. at
Amherst (1973), to appear in J. Pure Appl. Alg.

[6] - - Fuzzy Morphisms in Automata Theory, in: [60],
98 - 105

[7] ASENDORPF,J., Reduktion nichtdiskreter stochastischer
Automaten, Dipl.-Arb. Univ. Kiel 1973

[8] BAINBRIDGE,E.S., A Unified Minimal Realization Theory,
with Duality for Machines in a Hyperdoctrine
(Announcement of Results), Techn. Rep., Comp.and
Comm. Sci. Dept., Univ. of Michigan (June 1972)

[9] - - Addressed Machines and Duality, in: [60], 114-121

[10] BAUER,H., Wahrscheinlichkeitstheorie, Berlin 1968

[11] BENSON,D.B., An Abstract Machine Theory for Formal
Language Parsers, in: [60], 130 - 137

[12] BÖHLING,K.H. / INDERMARK,K., Endliche Automaten I ,
Mannheim 1969 (B.I. 703)

[13] BÖHLING,K.H. / SCHÜTT,D., Endliche Automaten II ,
Mannheim 1970 (B.I. 704)

[14] BRAUER,W., Zu den Grundlagen einer Theorie topologischer sequentieller Systeme und Automaten,
Berichte der GMD Bonn Nr. $\underline{31}$ (1970)

[15] - - Automates topologiques et ensembles
reconnaissables, Séminaire SCHÜTZENBERGER-LENTIN-
NIVAT Année 1969/70 no 18

[16] BRAUER,W., Automatentheorie, Stuttgart 1975

[17] BUDACH,L. / HOEHNKE,H.-J., Über eine einheitliche Begrün-
          dung der Automatentheorie, Seminarbericht
          1. Teil, HU Berlin 1969/70

[18] - -   Automaten und Funktoren (to appear)

[19] BURRONI,E., Algébres relatives à une loi distributive,
          C.R.Acad. Sci. Paris 276 (1973), Ser. A, 443-446

[20] CLAUS,V., Stochastische Automaten, Stuttgart 1971

[21] DAY,B., A Reflection Theorem for Closed Categories,
          J. Pure Appl. Alg. 2 (1972), 1 - 11

[22] DUGUNDJI,J., Topology, Boston 1966

[23] EHRIG,H., F-Morphismen (1972), Math. Nachr. 59 (1974),
          75 - 93

[24] - -   Kategorielle Theorie von Automaten, in: B.I.-
          Reihe Überblicke Mathematik 7 (1974) (editor:
          D. Laugwitz), 167 - 218

[25] - -   Automata Theory in Monoidal Categories, Proc. of
          the Tagung über Kategorien, Math. Forschungs-
          institut Oberwolfach 1972, 12 - 15

[26] - -   Kategorielle Theorie von Automaten und mehrdimen-
          sionalen formalen Sprachen (Zusammenfassender Be-
          richt), Forsch.-ber. 73-21 (1973), FB 20 TU Berlin

[27] EHRIG,H. / KIERMEIER,K.-D. / KREOWSKI,H.-J. / KÜHNEL,W.,
          Axiomatic Theory of Systems and Systematics,
          Forsch.-ber. 73 - 05 (1973), FB 20 der TU Berlin

[28] - -   Systematisierung der Automatentheorie, Seminar-
          bericht 1972/73, Forsch.-ber. 73 - 08 (1973),
          FB 20 der TU Berlin

[29] EHRIG,H. / KREOWSKI,H.-J., Power and Initial Automata in
          Pseudoclosed Categories, in: [60], 162 - 169

[30] - -   Systematic Approach of Reduction and Minimiza-
          tion in Automata and System Theory, Forsch.-ber.
          73 - 16 (1973), FB 20 der TU Berlin, to appear
          in revised version in: J. Comp. Syst. Sci.

[31] EHRIG,H. / KREOWSKI,H.-J. / PFENDER,M., Kategorielle The-
          orie der Reduktion, Minimierung und Äquivalenz
          von Automaten (1972), Math. Nachr. 59 (1974),
          105 - 124

[32] EHRIG,H. / KÜHNEL,W., Topological Automata, to appear in:
        Rev. Franç. d'Autom., d'Inf. et de Rech. Opér.

[33] EHRIG,H. / PFENDER,M. u.a., Kategorien und Automaten,
        Berlin - New York 1972

[34] EILENBERG,S., Automata, Languages and Machines, Vol. A ,
        New York - London 1974

[35] EILENBERG,S. / WRIGHT,J.B., Automata in General Algebras,
        Inform. and Contr. $\underline{11}$ (1967), 452 - 470

[36] ELGOT,C.C., The Common Algebraic Structure of Exit-
        Automata and Machines, Comp. $\underline{6}$ (1971), 349 - 370

[37] GÉCSEG,F. / PEAK,I., Algebraic Theory of Automata,
        Budapest 1972

[38] GINSBURG,S., An Introduction to Mathematical Machine
        Theory, Reading (Mass.) 1962

[39] GINZBURG,A., Algebraic Theory of Automata, New York 1968

[40] GIVEON,Y., Transparent Categories and Categories of
        Transition Systems, in: Proc. Conf. Cat. Alg.,
        La Jolla 1965

[41] GIVEON,Y. / ZALCSTEIN,Y., Algebraic Structures in Linear
        Systems Theory, J. Comp. Syst. Sci. $\underline{4}$ (1970),
        539 - 556

[42] GOGUEN,J.A.jr., L-Fuzzy Sets, J. Math. Anal. and Appl.
        $\underline{18}$ (1967), 145 - 174

[43] - -   Mathematical Representation of Hierarchically
        Organized Systems, Global Systems Dynamics, Int.
        Symp. Charlotteville 1969, 112 - 128

[44] - -   Discrete-Time-Machines in Closed Monoidal Cate-
        gories I , Quarterly Rep. no. 30, Inst. f. Comp.
        Res., Univ. of Chicago (1971), condensed version
        in: Bull. AMS $\underline{78}$ (1972), 777 - 783

[45] - -   Systems and Minimal Realization, Proc. of the
        1971 IEEE Conf. on Decision and Control,
        Miami Beach (1971), 42 - 46

[46] - -   Realization is Universal, Math. Syst. Theory
        vol. 6 no. 4 (1973), 359 - 374

[47] GROTEMEYER,K.P., Topologie, Mannheim 1969 (B.I. 836)

[48] - -   Lineare Algebra, Mannheim 1970 (B.I. 732)

[49] HARTMANIS,J. / STEARNS,R.E., Algebraic Structure Theory
       of Sequential Machines, Englewood Cliffs 1966

[50] HERRLICH,H. / STRECKER,G., Category Theory, Boston 1973

[51] HINDERER,K., Grundbegriffe der Wahrscheinlichkeits-
       theorie, Berlin - Heidelberg - New York 1972

[52] HOTZ,G., Übertragung automatentheoretischer Sätze auf
       Chomsky-Sprachen, Comp. $\underline{4}$ (1969), 30 - 42

[53] HOTZ,G. / CLAUS,V., Automatentheorie und Formale Spra-
       chen III , Mannheim 1971 (B.I. 823)

[54] HOTZ,G. / WALTER,H., Automatentheorie und Formale Spra-
       chen II , Mannheim 1970 (B.I. 822)

[55] HU,S.T., Mathematical Theory of Switching Circuits and
       Automata, Berkeley 1968

[56] KALMAN,R.E. / FALB,P.L. / ARBIB,M.A., Topics in Mathemat-
       ical System Theory, New York 1969

[57] KREOWSKI,H.-J., Automaten in pseudoabgeschlossenen
       Kategorien, Dipl.-Arb. TU Berlin 1974

[58] LINTON,F.E.J., Coequalizers in Categories of Algebras,
       in: Seminar on Triples and Categorical Homology
       Theory, Berlin - Heidelberg - New York 1969
       (Lecture Notes in Mathematics 80)

[59] MacLANE,S., Categories for the Working Mathematician,
       New York - Heidelberg - Berlin 1972

[60] MANES,E.G. (ed.), Proceedings of the 1. Intern. Symp.:
       Category Theory Applied to Computation and Con-
       trol, publ. by the Math. Dept. and the Dept. of
       Comp. and Inform. Sci., University of Mass.
       at Amherst 1974

[61] MESEGUER,J. / SOLS,I., Automata in Semimodule Categories,
       in: [60], 196 - 203

[62] NELSON,R.J., Introduction to Automata, New York 1968

[63] PAREIGIS,B., Categories and Functors, New York - London
       1970

[64] PFENDER,M., Kongruenzen, Konstruktion von Limiten und
       Cokernen und algebraische Kategorien, Diss.
       TU Berlin 1971

[65] PFENDER,M., Universal Algebra in S-monoidal Categories,
        Preprint TU Berlin 1973, to appear in:
        Algebra-Berichte Math. Inst. der Univ. München

[66] POHL,H.J., Über die Reduzierung der Anzahl von Eingabe-
        signalen von Automaten, Zeitschrift für math.
        Logik und Grundlagen d. Math. $\underline{11}$ (1968), 93 - 96

[67] - -   Ein Ansatz zur Theorie topologischer Automaten,
        Diss. HU Berlin 1971, to appear in EIK

[68] PUMPLÜN,D., Universelle und spezielle Probleme,
        Math. Ann. $\underline{198}$ (1972), 131 - 146

[69] REUSCH,B., Lineare Automaten, Mannheim 1969 (B.I. 708)

[70] SCHMITT,A., Zur Theorie der nichtdeterministischen und
        unvollständigen Automaten, Comp. $\underline{4}$ (1969), 56-74

[71] SCHUBERT,H., Categories, Berlin - Heidelberg - New York
        1972

[72] SEMADENI,Z., On Logical Educational Material, Categories
        and Automata, Queen's Math. Preprints no.
        1972 - 38 , Kingston

[73] SEMINAIRE BANACH, Berlin - Heidelberg - New York 1972
        (Lecture Notes in Mathematics 277)

[74] SPANIER,E., Quasi-Topologies, Duke Math. J. $\underline{30}$ (1963),1-14

[75] STARKE,P.H., Abstract Automata, Amsterdam - London 1972

[76] - -   Allgemeine Probleme und Methoden in der Automa-
        tentheorie, EIK $\underline{8}$ (1972), 489 - 517

[77] STEENROD,N.E., A Convenient Category of Topological
        Spaces, Mich. Math. J. $\underline{14}$ (1967), 133 - 152

[78] VALK,R., The Use of Metric and Uniform Spaces for the
        Formalization of Behavioral Proximity of States,
        in: 1. GI-Fachtagung über Automatentheorie und
        Formale Sprachen 1973, 116 - 122
        (Lecture Notes in Computer Science 2)

[79] WIWEGER,A., On Coproducts of Automata, Bull. Acad.
        Polonaise Sci. $\underline{21}$ (1973), 753 - 758

[80] ZADEH,L.A., Fuzzy Sets, Inform. and Contr. $\underline{8}$ (1965),
        338 - 353

Subject Index

Teubner Studienbücher Fortsetzung

Mathematik Fortsetzung

Stummel/Hainer: **Praktische Mathematik**
299 Seiten. DM 26,80

Topsøe: **Informationstheorie**
Eine Einführung. 88 Seiten. DM 11,80

Witting: **Mathematische Statistik**
Eine Einführung in die Theorie und Methoden
2. Aufl. 223 Seiten. DM 24,– (LAMM)

Physik   Elektrotechnik

Bourne/Kendall: **Vektoranalysis**
227 Seiten. DM 16,80

Daniel: **Beschleuniger**
215 Seiten. DM 22,–

Großmann: **Mathematischer Einführungskurs für die Physik**
264 Seiten. DM 22,80

Heber/Weber: **Grundlagen der Quantenphysik**
Band 1: Quantenmechanik. VI, 158 Seiten. DM 13,80
Band 2: Quantenfeldtheorie. VI, 178 Seiten. DM 14,80
(Vertrieb nur in der BRD und West-Berlin)

Kneubühl: **Repetitorium der Physik**
ca. 600 Seiten. DM 26,80

Lautz: **Elektromagnetische Felder**
Ein einführendes Lehrbuch. 180 Seiten. DM 17,80

Mayer-Kuckuck: **Physik der Atomkerne**
Eine Einführung. 2. Aufl. 288 Seiten. DM 19,80

Walcher: **Praktikum der Physik**
3. Aufl. 384 Seiten. DM 24,80

Mechanik

Becker: **Technische Strömungslehre**
Eine Einführung in die Grundlagen und technischen Anwendungen
der Strömungsmechanik. 3. Aufl. 144 Seiten. DM 11,80

Becker/Piltz: **Übungen zur Technischen Strömungslehre**
120 Seiten. DM 10,80

Magnus: **Schwingungen**
Eine Einführung in die theoretische Behandlung von Schwingungs-
problemen. 2. Aufl. 251 Seiten. DM 18,80 (LAMM)

Magnus/Müller: **Grundlagen der Technischen Mechanik**
300 Seiten. DM 24,– (LAMM)

Müller/Magnus: **Übungen zur Technischen Mechanik**
292 Seiten. DM 24,– (LAMM)

Wieghardt: **Theoretische Strömungslehre**
Eine Einführung. 2. Aufl. 237 Seiten. DM 24,– (LAMM)

Preisänderungen vorbehalten